The Urban Politics of Policy Failu

This book contributes to debates in geography and urban studies by analysing the spatial dimensions and politics of urban policy failure. Attention is most often paid to successful urban policies. Policymakers go to great lengths to emulate success by importing policy 'models', implementing best practices, or pursuing 'silver bullet' solutions. Yet, stories of failure are at least as common as those of success. Some policies fail to launch in the first place. Others struggle to deliver their goals. Many collapse under the weight of poor administration, insufficient funding, or political opposition.

This book establishes a vocabulary and set of analytical approaches for researching the spatial dynamics and impacts of urban policy failure. With a geographically diverse set of cases, the authors explore topics including policy (im)mobility, urban policy experiments, and governance initiatives ranging from sustainability to housing to public health, across Europe, North America, and Asia.

The chapters in this book were originally published as a special issue of the journal *Urban Geography*.

John Lauermann is Assistant Professor of Geography at Medgar Evers College and The Graduate Center of the City University of New York. He is an urban geographer interested in the planning and socio-spatial impacts of large real estate developments.

Cristina Temenos is Senior Lecturer in Urban Geography and UKRI Future Leaders Fellow in Urban Crisis at the University of Manchester. She is an urban geographer interested in the politics and social processes of making and moving urban policy.

The Urban Politics of Policy Failure

Edited by
John Lauermann and Cristina Temenos

Routledge
Taylor & Francis Group

LONDON AND NEW YORK

First published 2023
by Routledge
4 Park Square, Milton Park, Abingdon, Oxon OX14 4RN

and by Routledge
605 Third Avenue, New York, NY 10158

Routledge is an imprint of the Taylor & Francis Group, an informa business

© 2023 Taylor & Francis

British Library Cataloguing in Publication Data
A catalogue record for this book is available from the British Library

ISBN13: 978-1-032-26858-3 (hbk)
ISBN13: 978-1-032-26859-0 (pbk)
ISBN13: 978-1-003-29024-7 (ebk)

DOI: 10.4324/9781003290247

Typeset in Minion Pro
by Newgen Publishing UK

Publisher's Note
The publisher accepts responsibility for any inconsistencies that may have arisen during the conversion of this book from journal articles to book chapters, namely the inclusion of journal terminology.

Disclaimer
Every effort has been made to contact copyright holders for their permission to reprint material in this book. The publishers would be grateful to hear from any copyright holder who is not here acknowledged and will undertake to rectify any errors or omissions in future editions of this book.

Contents

Citation Information

The chapters in this book were originally published in *Urban Geography*, volume 41, issue 09 (2020). When citing this material, please use the original page numbering for each article, as follows:

Introduction
The urban politics of policy failure
Cristina Temenos and John Lauermann
Urban Geography, volume 41, issue 09 (2020), pp. 1109–1118

Chapter 1
Going bust two ways? Epistemic communities and the study of urban policy failure
Mark Davidson
Urban Geography, volume 41, issue 09 (2020), pp. 1119–1138

Chapter 2
Policy-failing: a repealed right to shelter
Katie J. Wells
Urban Geography, volume 41, issue 09 (2020), pp. 1139–1157

Chapter 3
Urban policy (im)mobilities and refractory policy lessons: experimenting with the sustainability fix
Aida Nciri and Anthony Levenda
Urban Geography, volume 41, issue 09 (2020), pp. 1158–1178

Chapter 4
Beyond failure: the generative effects of unsuccessful proposals for Supervised Drug Consumption Sites (SCS) in Melbourne, Australia
Tom Baker and Eugene McCann
Urban Geography, volume 41, issue 09 (2020), pp. 1179–1197

For any permission-related enquiries please visit:
www.tandfonline.com/page/help/permissions

Notes on Contributors

Tom Baker, School of Environment, University of Auckland, Auckland, New Zealand.

Rachel Bok, Department of Geography, University of British Columbia, Vancouver, BC, Canada.

Christine Bonnin, School of Geography, University College Dublin, Belfield, Ireland.

Mark Davidson, School of Geography, Clark University, Worcester, MA, USA.

John Lauermann, Medgar Evers College and The Graduate Center, City University of New York, New York, NY, USA.

Anthony Levenda, Center for Climate Action and Sustainability, The Evergreen State College, Olympia, WA, USA.

Eugene McCann, Department of Geography, Simon Fraser University, Burnaby, Canada.

Niamh Moore-Cherry, School of Geography, University College Dublin, Dublin, Ireland.

Aida Nciri, Department of Geography, University of Calgary, Calgary, Alberta, Canada Laboratoire Techniques, Territoires et Sociétés (LATTS), Université Paris-Est, Marne-la-Vallée, France.

Cristina Temenos, Department of Geography, University of Manchester, Manchester, UK.

Katie J. Wells, Kalmanovitz Initiative for Labor and the Working Poor, Georgetown University, Washington, DC, US.

Acknowledgements

The editors would like to thank all the authors for their contributions, as well as all presenters in the sessions on the Urban Politics of Policy Failure at the 2017 American Association of Geographers conference. We would also like to thank Susan Moore for her support in editing the original Special Issue of *Urban Geography*, and Anveshi Gupta for her support in putting together this volume. Last but not least, we thank our families for their ongoing support, particularly Erin, Hannah, Nick and Paola.

Cristina Temenos would like to acknowledge that this work was supported by the Medical Research Council Future Leaders Fellowship [grant number MR/V02468X/1].

The urban politics of policy failure

Cristina Temenos and John Lauermann ⓘD

ABSTRACT
Growing attention to the urban politics of policy failure focuses on the ways in which policymaking is politicized before, during, and after moments of failure, and the effects of policy failure in urban political process. We argue three trends are apparent: (1) Policy failure is a relational process rather than an end state, emerging from long term interactions between political actors and governance institutions. (2) In urban politics the definitions of "failure" and "success" can be fluid, and rhetorically weaponized for political gain. (3) Policy failures can have generative effects after apolicy fails, for instance, by catalyzing funding or establishing institutional path dependencies. This collection contributes to urban geography by establishing avocabulary and set of analytical techniques for researching spatial aspects of policy failure. The authors in this special issue build an empirical and theoretical baseline for future research, illustrating diverse mechanisms and outcomes of urban policy failure.

Introduction

Much attention has been paid to urban governance success stories in recent years. Policymakers – and the scholars who analyze them – go to great lengths to emulate these success stories by importing models from elsewhere, experimenting with pilot programs, or implementing best practices. Yet stories of failure are at least as common as those of success. Indeed, the history of urban studies is littered with tales of "unurban urbanization" (Jacobs, 1961), "urban crisis" (Harvey, 1978), "great planning disasters" (Hall, 1980), "splintered" landscapes (Graham & Marvin, 2001), "dysfunctional urbanism" (McNeill, 2005), or the many reminders for policymakers to pay careful attention to "actually existing" urban conditions – including not only the success stories but also evidence of policy ineffectiveness and failure (Brenner & Theodore, 2002; Clarke, 2012; Eizenberg, 2012; Krueger & Agyeman, 2005; Shatkin, 2011; Shelton et al., 2015).

This special issue examines the urban politics of policy failure through the lens of geography and urban studies. This is an emerging literature drawing on a constellation of inter-related terms, including "differentiation, mutation, fragility, unraveling, instability, emergence, detour, redirection, reaction, rejection, de-activation, and absence" (Baker & McCann, This Issue, p. 7). In this literature, the term "policy failure" typically refers to policies that fail to launch: policies that were thwarted, canceled, stalled, or otherwise prevented from reaching full implementation. But the term can include other forms of failure as well, for instance, policies that "have failed to deliver on their promises," policies that produce "deleterious social outcomes," or instances in which "policy failure has not stymied policy mobility" (Davidson, This Issue, pp. 5–6). This special issue takes an eclectic approach to definitions. Indeed, the following articles work across many of the abovementioned definitions of failure. Our interest, rather, is analyzing the urban politics which surround policy failures in their diverse spatialities: the ways in which policy-making is politicized before, during, and after a moment of failure, and the effects of policy failure in the urban political process.

This collection contributes to urban geography by establishing a vocabulary and set of analytical approaches for researching policy failure. Collectively, the authors build an empirical baseline for future research, illustrating the mechanisms and outcomes of urban policy failure. They also identify key theoretical properties which inform spatial and spatio-temporal analysis of policy failure, properties which our introductory essay synthesizes in the following section.

Politicizing urban policy failure

In the urban geography literature, there is a growing sense that "failure matters" (Chang, 2017) in the analysis of urban space, its production, and its politics. Indeed, there is strong evidence that policy failure is endemic to urban governance in late capitalism, a form of creative destruction which facilitates entrepreneurial forms of governance. As Brenner et al. (2010, p. 333) put it,

> policy failure is not only central to the exploratory modus operandi of neoliberalization processes; it provides a further, powerful impetus for their accelerating proliferation and continual reinvention across sites and scales. Crucially, then, endemic policy failure has actually tended to spur further rounds of reform within broadly neoliberalized political and institutional parameters: it triggers the continuous reinvention of neoliberal policy reper-toires rather than their abandonment.

In this sense, failure is dialectically entwined with urban governance. As studies in this special issue suggest, it is an obvious, yet under-researched, component of the process of trial and error through which policy is made. Just as importantly for our purposes, policy failure is a highly politicized aspect of policymaking: a cautionary tale, a means of subversion, a form of protest, an opportunity to advance alternative agendas.

Several urban geography sub-literatures have begun to explore the systemic role of policy failure in capitalist urban governance. Entrepreneurial city researchers have pointed out that, by definition, entrepreneurial urban governance involves risk-taking investment by municipalities. Failed investments are an obvious potential outcome of such behavior, especially as cities involve themselves in complicated, speculative forms of financialization (Akers, 2015; Weber, 2015). Impacts which follow from failed entrepre-neurial investments include municipal bankruptcy (Peck, 2014), municipal austerity (Fuller, 2018), and "degrowth" politics (Béal et al., 2019; Schindler, 2016). A similar surge in policy failure research is taking place in the policy mobilities literature. Researchers have highlighted the significance of "policy immobilities", cases in which a policy fails to move despite efforts at policy transfer (Malone, 2019; Müller, 2015; Wood, 2019). They have also identified examples of "policy failure mobilities" (Lovell, 2019) in which negative case studies travel across cities as cautionary tales of worst practice to avoid. Indeed these forms of failure play an important role in policy learning, as supplements to the "success stories" which motivate policy circulation (Stein et al., 2017). Finally, research on experimental urban governance – and related conversations in urban science and technology studies – highlights that failure is a clear possibility in any experiment's design (Evans et al., 2016). Indeed, the potential for failure is part of the appeal in this trial-and-error form of governance (Evans, 2016), though failed experiments may lumber forward as a form of "Frankenstein urbanism" (Cugurullo, 2018) despite failing to pass their own experimental standards.

Three dimensions stand out in the analysis of urban policy failures, which we synthesize here and which the remaining articles of the issue analyze in detail. First, policy failure is a relational process, emerging from long term interactions between political actors and institutions. Any policymaking process contains definitive moments when a policy may fail, yet "the making of a policy may fail temporarily, repeatedly, or permanently. What demands attention is not the end product of botched governance efforts but the actual practices and conditional forces that create these moments of policyfailing." (Wells, 2014, p. 475). Full stop failure is only one of several possible outcomes in a policymaking process, as urban governance proceeds in stops and starts through parallel and overlapping forms of trial and error. The leadership and funding of political institutions matter greatly in the failure process, as do moments of institutional change associated with elections, legislation, and political or economic crises.

Several authors in this special issue explore the process of failure; how it is actually experienced and implemented. They analyze realities of failure across diverse geographic contexts and considers conceptual potentials and limits of different approaches. Wells (This Issue) analyses the ways in which policies are made to fail through a repealed "right to shelter" law in Washington D.C. She details the deliberate choices made in the process of ensuring that this policy would not realize its goals and argues that such actions and inactions expose the political nature of policy-making in ways that contradict established thinking on urban governance. She notes how: "The failing of the right to shelter was a protracted, laborious, and multi-sited political-geographic process. ... empirical data does not reveal a singular moment, site, or mechanism by which the right to shelter was made to fail. Instead the failing of the right to shelter emerged from an array of places of power, including the judiciary, the local executive and legislative branches, the media, and the streets." (pp 14–15) Likewise, Davidson (This Issue) analyzes the politics of municipal bankruptcy with a study of bankruptcy proceedings in Vallejo, California. He utilizes two different epistemological approaches, critical political economy and public policy, demonstrating that while failure is apparent in both analyses, explanatory functions are not always commensurable, and the stories told are shaped by preexisting understandings of failure. In a different vein, Bok (This Issue) explores the notion of failure by charting the failure of the previously proven Singapore model in China over time, highlighting the importance of state structures and political histories in how policy models are implemented in the present.

Second, in urban politics the definitions of "failure" and "success" are not fixed, and can be rhetorically weaponized for political gain. Writing on intellectual dualisms belying the policy mobilities literature, McCann and Ward (2015, p. 1) argue that failure/success is a particularly insidious construct, because

> neither success nor failure is absolute. One does not make sense without the other. Rather, success and failure are relationally constituted in politics and in policymaking. Studies of urban policy mobilities should, then, reflect critically on approaches to success/failure and their relational constitution, even as they simultaneously study the effects of their empirical separation and their reification in policymaking.

Similarly, Stein et al. (2017) highlight a "success bias" in urban politics (and urban scholarship), in the sense that success stories are touted widely while failures are ignored and forgotten. As rhetorical categories in urban politics, failure and success can be strategically leveraged depending on audience and agenda. The fickle nature of political opinion about development projects for example, demonstrates the shifting category of success and failure (Mattissek & Sturm, 2017). In their discussion of mega-projects, Holden et al. (2015, p. 451) note: "what counts as failure and as success in the work of city-building will shift, depending on what actors do and how they talk about it, and on how well these actions and justifications hold up to public challenges about the true character of a successful city." This ambiguity does significant political work to support or attack policy. Nonetheless, it presents analytical challenges for policy failure researchers, who must interpret the positionality of policy and policy actors vis-à-vis the "failure".

In this special issue, Moore-Cherry and Bonnin (This Issue) argue that definitions of success and failure vary based on multiple, competing temporalities. In their study of the Moore Street Market, the oldest surviving street market in Dublin, they demonstrate that temporal framings and competing timelines can shift the definitions of failure and success within urban governance and planning processes. "Whether the recent past and future of Moore Street will be seen as a 'success' or 'failure' depends on the temporal framings we privilege. Whether we frame our interpretation through the eyes of the traders, campaigners, central government, local authority planners, or developers in the short, medium or long-term will provide a very different reading of the street." (17–18) Furthermore, they show how planning initiatives are also often powerless at times in dictating the timeframes within which development can occur. Likewise, Baker and McCann (This Issue), in their study of repeatedly failed policy proposals to institute a supervised consumption site for people using illicit drugs in Melbourne, demonstrate the value of a longitudinal view in evaluating success and failure. It shifts focus to definitions of how failure is constituted, and whether potential "successes" such as network building and subject literacy might be found out of policy failures.

Such successes however are always in flux. Nciri and Levenda (This Issue) demonstrate that lesson-drawing often only happens when policy lessons are rendered ideologically legible, based on their analysis of the failure of district heating, a low-carbon technology, to catch on in the Calgary Metropolitan Region. The importance of discursive strategies has long been a focus of policy mobilities work (Temenos & McCann, 2012). Nciri and Levanda build on this work by analyzing the strategies that politicians and policy makers used to ensure that the district heating scheme, which was successful in meeting its sustainability targets, was nonetheless constructed as a failure through its challenge to profit margins of private development. They note how the "politics of urban experiments are deeply intertwined with the social and power-laden construction of success and failure (and who defines them as such) ... actors can mobilize lessons from experimentation to slow down or block sustainable transition." (2) Similarly, Bok (This Issue) highlights the role that geopolitical anxiety over sovereignty plays in the interpretations of the failure of the Singapore model in Chinese cities. She notes "The instances of 'failure' experienced by Singaporean officials in the SSTEC [Sino-Singapore Tianjin Eco-city] capture not only the ambivalence of 'success' itself but, more significantly, how what is considered as 'failure' transcends individual projects and is in reality reflective of broader (geo)political circumstances, culminating in the 'hollowing out' of a signifier of success." (21)

Third, there is ample evidence that policy failures can have generative effects after a policy fails. Chang (2017) shows, for instance, that repeated failures to build eco-cities in China generate path dependencies in urban policy despite their failure, by establishing planning procedures and repeatedly hiring particular groups of consultants. Lauermann (2016) shows, likewise, that cities' failed bids to host mega-events often contribute to infrastructure investment, as more viable sub-components of the plans are pushed forward even though the grand plan is abandoned. Tracing these generative effects often requires digging into the empirical details of contracts, intra-institutional decision-making, and the career trajectories of political and policy actors. Even with detailed data, establishing clear causal relationships between a policy failure and a generative effect may be difficult, given loss of institutional memory over time and the political recriminations that may follow from failures. Nonetheless, it is clear that failures shape institutional and political decisions in subtle-yet-profound ways.

In the special issue, Baker and McCann (This Issue) trace the multiple ways in which the geography of absence was still able to alter the parameters of political debates. In their longitudinal study of the repeated attempts to open a supervised drug consumption site in Melbourne, their work traces the series of failures through digging into old reports, proposals and interviews with advocates and medical doctors involved in the attempts. Despite the opposition of local government to establishing the health service they argue that "failure was also generative. It influenced the individual careers of harm reduction advocates and experts, and supported the further development of local, national, and global networks of policy and activism knowledge and a longstanding commitment to the cause of harm reduction that has influenced the character and practice of drug policy, drug treatment, and drug activism more generally." (15) They go on to show that this had effects that went beyond the urban site, contributing to drug policy literacy at regional, national, and international scales. Failure in this sense was not absolute; rather, it co-produced other forms of policy action across multiple spatialities.

Next steps

The question of "why study failure?" is one that Davidson (This Issue) explicitly addresses in this special issue, and it is a question that pervades all the articles herein. In looking forward, it is instructive to look at the present. We write this introduction in the midst of a global lockdown intended to stem the spread of COVID-19. We find ourselves in this position due to a series of failures. Failure of individual actors to heed the warnings of experts, failure of governments at all levels to act quickly to implement appropriate testing and isolation measures, failure of governments and markets to address critical shortages of appropriate protective gear and medical equipment. These failures matter because they have cost hundreds of thousands of lives, they matter because domestic violence rates have doubled since the outbreak (Townsend, 2020), they matter because livelihoods have been destroyed. Furthermore, the multiple failures leading up to and ongoing within the pandemic serve to expose long-standing policy failures: structural racism, housing precarity, environmental degradation, entrenched poverty, and precarious work among many others. A critical attention to the politics of failure, then, works to re-center power within analyses of urban policy, laying bare the multiple and ongoing failures in ways that can help us to imagine alternative policy interventions and, more radically, alternate forms of governance, some of which may prioritize social reproduction over or alongside financialization and economic growth.

In geography, analysis of policy failure is more likely to be interpretive than diagnostic. This contrasts with the policy failure literature in political science and public policy, which tends to focus on diagnosing the factors that contribute to success and failure (Dunlop, 2017; Howlett et al., 2015; McConnell, 2010). In a burgeoning literature like that on policy failures, there is ample room for discipline-specific specialization. Nonetheless, as Davidson (This Issue, p. 7) argues, geographers must "face the question of what we want to know about policy failures." In much of the urban geography literature on the topic, the answer thus far seems to be interpretive insight into broader political economic processes shaping urban governance. That is, policy failure *reveals*. Analyzing episodes of failure can help explain the often inscrutable gap between what policy actors say and what they actually do, telling us much about the internal dynamics and dysfunctions of governing institutions. But future research can expand the scope of analysis to consider a number of factors, including temporality, relationality, and generative effects.

Temporality, as several authors in this collection have highlighted, is a key area for critical geographies of failure. Longitudinal or historic methods are often necessary to track the impact of policy failure. Processes, decisions, policy implementation and effect necessarily play out over various time scales. The relationship of time to the geographies of failure can serve to demonstrate how prior failures become enmeshed in building urban futures. Furthermore, analysis of how temporality affects urban geographies of policy failure expands the horizon of this field beyond attention to neoliberal "fast policy" mobility (Peck & Theodore, 2015). Whether through their incorporation into new policy objectives, shaping the discourses of new perceptions of possibility, foreclosure of previous debates, or allocation of funds, policy failures are felt materially through the urban built environment ranging from iconic architectures of grandiose yet misguided initiatives to mundane and dysfunctional infrastructures that are quietly built into the fabric of the city.

The argument for a temporal turn is also an argument for a relational one (Massey, 2005). The relational turn in urban geography is not new, yet the renewed interest in this epistemological thread brings new actors into debates and analyses of urban policy failure, not only focusing on policy makers (McCann & Ward, 2011; Ward, 2018), but also activists (Temenos, 2017; Lauermann & Vogelpohl, 2019), everyday actors (Baker et al., 2020; Jacobs & Lees, 2013), and planners and consultants (Colven, 2020; Larner & Laurie, 2010; Rapoport, 2015; Vogelpohl, 2018). Expanding the scope of study to encompass new relationships can broaden and enrich the spatial analysis of urban geographies of policy, taking into account different power structures that make up local contexts and expanding the scope of possibility for urban politics.

Finally, the generative effects of failed policy are an important area of further research. Looking to what comes after failure is a useful way to understand urban politics, whether it be to understand the persistence of neoliberal policies that exacerbate inequality (Wells, This Issue; Peck, 2014), or to think though progressive or even radical urban futures (Baker & McCann, This Issue; Russell, 2019; Sutton, 2019). New municipalism, for example, is a governance movement rising in resistance to austerity urbanism and the failures of the state to work within the existing social contract (Russell, 2019; Thompson, 2020). Widening inequality, increasing poverty, housing and labor insecurity, negative health outcomes: these are all areas where urban governments have failed to achieve meaningful change and the new municipalism movement looks to use urban governance as a strategic space for progressive or transformative politics. The afterlives of failed policies and their transformative political effects then have the potential to provide answersto Davidson's (This Issue) question of the utility of studying the urban politics of policy failure.

Acknowledgments

The authors would like to thank all participants in the 2017 AAG conference sessions on the Urban Politics of Policy Failure which inspired this Special Issue.

Disclosure statement

No potential conflict of interest was reported by the authors.

ORCID

John Lauermann ⓘ http://orcid.org/0000-0001-9114-3864

References

Akers, J. (2015). Emerging market city. *Environment and Planning A: Economy and Space*, *47*(9), 1842–1858. https://doi.org/10.1177/0308518X15604969

Baker, T., & McCann, E. (This Issue). Beyond failure: The generative effects of unsuccessful proposals for supervised drug consumption sites (SCS) in Melbourne, Australia. *Urban Geography*. http://dx.doi.org/10.1080/02723638.2018.1500254

Baker, T., McCann, E., & Temenos, C. (2020). Into the ordinary: Non-elite actors and the mobility of harm reduction policies. *Policy and Society*, *39*(1), 129–145. https://doi.org/10.1080/14494035.2019.1626079

Béal, V., Fol, S., Miot, Y., & Rousseau, M. (2019). Varieties of right-sizing strategies: Comparing degrowth coalitions in French shrinking cities. *Urban Geography*, *40*(2), 192–214. https://doi.org/10.1080/02723638.2017.1332927

Brenner, N., Peck, J., & Theodore, N. (2010). After neoliberalization? *Globalizations*, *7*(3), 327–345. https://doi.org/10.1080/14747731003669669

Brenner, N., & Theodore, N. (2002). Cities and the geographies of "actually existing neoliberalism". *Antipode*, *34*(3), 349–379. https://doi.org/10.1111/1467-8330.00246

Chang, I.-C. C. (2017). Failure matters: Reassembling eco-urbanism in a globalizing China. *Environment and Planning A: Economy and Space*, *49*(8), 1719–1742. https://doi.org/10.1177/0308518X16685092

Clarke, N. (2012). Actually existing comparative urbanism: Imitation and cosmopolitanism in North-south interurban partnerships. *Urban Geography, 33*(6), 796–815. https://doi.org/10.2747/0272-3638.33.6.796

Colven, E. (2020). Thinking beyond success and failure: Dutch water expertise and friction in postcolonial Jakarta. *Environment and Planning C: Politics and Space* 38(6): 961-979. 2399654420911947. https://doi.org/10.1177/2399654420911947

Cugurullo, F. (2018). Exposing smart cities and eco-cities: Frankenstein urbanism and the sustainability challenges of the experimental city. *Environment and Planning A: Economy and Space, 50* (1), 73–92. https://doi.org/10.1177/0308518X17738535

Davidson, M. (This Issue). Going bust two ways? Epistemic communities and the study of urban policy failure. *Urban Geography.* http://dx.doi.org/10.1080/02723638.2019.1621122

Dunlop, C. A. (2017). Policy learning and policy failure: Definitions, dimensions and intersections. *Policy & Politics, 45*(1), 3–18. https://doi.org/10.1332/030557316X14824871742750

Eizenberg, E. (2012). Actually existing commons: Three moments of space of community gardens in New York City. *Antipode, 44*(3), 764–782. https://doi.org/10.1111/j.1467-8330.2011.00892.x

Evans, J. (2016). Trials and tribulations: Problematizing the city through/as urban experimentation. *Geography Compass, 10*(10), 429–443. https://doi.org/10.1111/gec3.12280

Evans, J., Karvonen, A., & Raven, R. (Eds.). (2016). *The experimental city.* Routledge.

Fuller, C. (2018). Entrepreneurial urbanism, austerity and economic governance. *Cambridge Journal of Regions, Economy and Society, 11*(3), 565–585. https://doi.org/10.1093/cjres/rsy023

Graham, S., & Marvin, S. (2001). *Splintering urbanism: Networked infrastructures, technological mobilities and the urban condition.* Routledge.

Hall, P. (1980). *Great planning disasters.* University of California Press.

Harvey, D. (1978). The urban process under capitalism: A framework for analysis. *International Journal of Urban and Regional Research, 2*(1–3), 101–131. https://doi.org/10.1111/j.1468-2427.1978.tb00738.x

Holden, M., Scerri, A., & Esfahani, A. H. (2015). Justifying redevelopment 'failures' within urban 'success stories': Dispute, compromise, and a new test of urbanity. *International Journal of Urban and Regional Research, 39*(3), 451–470. https://doi.org/10.1111/1468-2427.12182

Howlett, M., Ramesh, M., & Wu, X. (2015). Understanding the persistence of policy failures: The role of politics, governance and uncertainty. *Public Policy and Administration, 30*(3–4), 209–220. https://doi.org/10.1177/0952076715593139

Jacobs, J. (1961). *The death and life of great American cities.* Vintage Books.

Jacobs, J. M., & Lees, L. (2013). Defensible space on the move: Revisiting the urban geography of Alice Coleman. *International Journal of Urban and Regional Research, 37*(5), 1559–1583. https://doi.org/10.1111/1468-2427.12047

Krueger, R., & Agyeman, J. (2005). Sustainability schizophrenia or "actually existing sustainabilities?" toward a broader understanding of the politics and promise of local sustainability in the US. *Geoforum, 36*(4), 410–417. http://dx.doi.org/10.1016/j.geoforum.2004.07.005

Larner, W., & Laurie, N. (2010). Travelling technocrats, embodied knowledges: Globalising privatisation in telecoms and water. *Geoforum, 41*(2), 218–226. https://doi.org/10.1016/j.geoforum.2009.11.005

Lauermann, J. (2016). Temporary projects, durable outcomes: Urban development through failed Olympic bids? *Urban Studies, 53*(9), 1885–1901. http://dx.doi.org/10.1177/0042098015585460

Lauermann, J., & Vogelpohl, A. (2019). Fast activism: Resisting mobile policies. *Antipode, 51*(4), 1231–1250. https://doi.org/10.1111/anti.12538

Lovell, H. (2019). Policy failure mobilities. *Progress in Human Geography, 43*(1), 46–63. https://doi.org/10.1177/0309132517734074

Malone, A. (2019). (Im)mobile and (Un)successful? A policy mobilities approach to New Orleans's residential security taxing districts. *Environment and Planning C: Politics and Space, 37*(1), 102–118. https://doi.org/10.1177/2399654418779822

Massey, D. (2005). *For space.* Sage.

Mattissek, A., & Sturm, C. (2017). How to make them walk the talk: Governing the implementation of energy and climate policies into local practices. *Geographica Helvetica*, 72(1), 123. https://doi.org/10.5194/gh-72-123-2017

McCann, E., & Ward, K. (Eds.). (2011). *Mobile urbanism: Cities and policymaking in the global age* (Vol. 17). U of Minnesota Press.

McCann, E., & Ward, K. (2015). Thinking through dualisms in urban policy mobilities. *International Journal of Urban and Regional Research*, 39(4), 828–830. https://doi.org/10.1111/1468-2427.12254

McConnell, A. (2010). Policy success, policy failure and grey areas in-between. *Journal of Public Policy*, 30(3), 345–362. https://doi.org/10.1017/S0143814X10000152

McNeill, D. (2005). Dysfunctional urbanism. *International Journal of Urban and Regional Research*, 29(1), 201–204. https://doi.org/10.1111/j.1468-2427.2005.00579_2.x

Moore-Cherry, N., & Bonnin, C. (This Issue). Playing with time in Moore Street, Dublin: Urban redevelopment, temporal politics and the governance of space-time. *Urban Geography*. http://dx.doi.org/10.1080/02723638.2018.1429767

Müller, M. (2015). (Im-)mobile policies: Why sustainability went wrong in the 2014 Olympics in Sochi. *European Urban and Regional Studies*, 22(2), 191–209. https://doi.org/10.1177/0969776414523801

Nciri, A., & Levenda, A. (This Issue). Urban policy (im) mobilities and refractory policy lessons: Experimenting with the sustainability fix. *Urban Geography*. http://dx.doi.org/10.1080/02723638.2019.1575154

Peck, J. (2014). Pushing austerity: State failure, municipal bankruptcy and the crises of fiscal federalism in the USA. *Cambridge Journal of Regions, Economy and Society*, 7(1), 17–44. https://doi.org/10.1093/cjres/rst018

Peck, J., & Theodore, N. (2015). *Fast policy: Experimental statecraft at the thresholds of neoliberalism*. U of Minnesota Press.

Rapoport, E. (2015). Globalising sustainable urbanism: The role of international masterplanners. *Area*, 47(2), 110–115. https://doi.org/10.1111/area.12079

Russell, B. (2019). Beyond the local trap: New municipalism and the rise of the fearless cities. *Antipode*, 51(3), 989–1010. https://doi.org/10.1111/anti.12520

Schindler, S. (2016). Detroit after bankruptcy: A case of degrowth machine politics. *Urban Studies*, 53(4), 818–836. https://doi.org/10.1177/0042098014563485

Shatkin, G. (2011). Coping with actually existing urbanisms: The real politics of planning in the global era. *Planning Theory*, 10(1), 79–87. https://doi.org/10.1177/1473095210386068

Shelton, T., Zook, M., & Wiig, A. (2015). The 'actually existing smart city'. *Cambridge Journal of Regions, Economy and Society*, 8(1), 13–25. https://doi.org/10.1093/cjres/rsu026

Stein, C., Michel, B., Glasze, G., & Pütz, R. (2017). Learning from failed policy mobilities: Contradictions, resistances and unintended outcomes in the transfer of "Business Improvement Districts" to Germany. *European Urban and Regional Studies*, 24(1), 35–49. https://doi.org/10.1177/0969776415596797

Sutton, S. A. (2019). Cooperative cities: Municipal support for worker cooperatives in the United States. *Journal of Urban Affairs*, 41(8), 1081–1102. https://doi.org/10.1080/07352166.2019.1584531

Temenos, C. (2017). Everyday proper politics: Rereading the post-political through mobilities of drug policy activism. *Transactions of the Institute of British Geographers*, 42(4), 584–596. https://doi.org/10.1111/tran.12192

Temenos, C., & McCann, E. (2012). The local politics of policy mobility: Learning, persuasion, and the production of a municipal sustainability fix. *Environment and Planning A*, 44(6), 1389–1406. https://doi.org/10.1068/a44314

Thompson, M. (2020). What's so new about New Municipalism? *Progress in Human Geography*, 030913252090948. http://dx.doi.org/0309132520909480

Townsend, M. 12 April 2020, *Revealed: Surge in domestic violence during Covid-19 crisis*. The Guardian. Retrieved April 16, 2020, from https://www.theguardian.com/society/2020/apr/12/domestic-violence-surges-seven-hundred-per-cent-uk-coronavirus

Vogelpohl, A. (2018). Consulting as a threat to local democracy? Flexible management consultants, pacified citizens, and political tactics of strategic development in German cities. *Urban Geography*, *39*(9), 1345–1365. https://doi.org/10.1080/02723638.2018.1452872

Ward, K. (2018). Urban redevelopment policies on the move: Rethinking the geographies of comparison, exchange and learning. *International Journal of Urban and Regional Research*, *42* (4), 666–683. https://doi.org/10.1111/1468-2427.12604

Weber, R. (2015). *From boom to bubble: how finance built the New Chicago*. University of Chicago Press.

Wells, K. (2014). Policyfailing: The case of public property disposal in Washington, D.C. *ACME*, *13*(3), 473–494.

Wells, K. J. (This Issue). Policy-failing: A repealed right to shelter. *Urban Geography*, 1–19. https://doi.org/10.1080/02723638.2019.1598733

Wood, A. (2019). Circulating planning ideas from the metropole to the colonies: Understanding South Africa's segregated cities through policy mobilities. *Singapore Journal of Tropical Geography*, *40*(2), 257–271. https://doi.org/10.1111/sjtg.12273

Going bust two ways? Epistemic communities and the study of urban policy failure

Mark Davidson

ABSTRACT

Urban geographers are becoming more concerned with "policy failure". This raises questions about how "policy failure" should be conceptualized. The public policy literature, with its detailed classifications and categorizations of policy failure, is an obvious potential resource for urban geographers. However, supplementing predominant urban geographical analysis with public policy frameworks presents significant epistemological challenges. The literatures belong to different disciplinary traditions, making a simple combination of the two difficult. To demonstrate, the paper presents two contrasting accounts of a recent case of "policy failure": the 2008 bankruptcy of the City of Vallejo, California. The accounts are distinguished by their epistemological orientations, one based in theoretical explanation (geography) and the other concerned with practical explanation (public policy). When we acknowledge these epistemological differences, we are forced to assess the limits to synthesizing different types of urban policy failure analysis. In conclusion, the paper discusses the pragmatic approach to epistemological choice.

"In theory, there is no difference between theory and practice. In practice there is." – *variously attributed.[1]*

Introduction

The above adage summarizes the problem urban geographers face in their consideration of "policy failure" (Jacobs, 2012; Lovell, 2019; McCann & Ward, 2015; Webber, 2015). Theorizing about urban policy making and understanding how policies are practiced are prospectively not incompatible. But, in practice, they often are. The reason, I argue, is that undertaking these tasks usually involve employing different types of reasoning; theoretical and practical. In urban geography, a longstanding concern with how urban policies reflect dominant ideologies has meant the intricacies of policymaking and implementation are often not the focus of collective epistemological projects. Policies become successful in their realization of ideology: stadia are built, waterfronts reinvested, neighborhoods gentrified (Lovell,

2019). Urban geographers therefore tend to write critical accounts of realization, not detailed studies of policy formulation and implementation (e.g. Newman & Ashton, 2004).

Examining policy failure represents a different challenge. Often a policy fails precisely because it does not create change (Jacobs, 2012). Or a policy partially fails, making it difficult to assign responsibility for complex urban changes (Clarke, 2012). These challenges have led some to adopt conceptual and methodological approaches from other (sub)disciplines. For example, Lovell (2019) has borrowed from political science, science-and-technology studies (STS), and economic geography to study policy mobilities failure. Lovell argues these fields offer the conceptual and methodologies resources required for urban geographers to understand policy failure:

> "political science, economic geography and STS scholarship provide direction and insight for the study of policy failure mobilities ... the way forward involves not just better methodological balance and attention across policy successes and failures, but also further conceptual development within policy mobilities research." (ibid. 13)

These arguments (Lovell, 2019; also see, 2016, 2017b) for interdisciplinarity are convincing. However, there are significant epistemological challenges that come along with this project.

Some of these epistemological challenges are reflected in prior debates about policy relevance (Hamnett, 2003; Markusen, 1999). These debates were instigated by claims that urban geography had become progressively disconnected from the world of policy-making (Dorling & Shaw, 2002; Martin, 2001). The result being that geographical knowledge is now rarely heard or understood by policy-makers (Dorling & Shaw, 2002). Imrie (2004) developed a more sympathetic critique, arguing that the epistemological orientation of urban geography has often made it inapplicable to the world of "evidence-based decision-making":

> " ... urban geographers ought not to be defensive about their subject, or necessarily apologetic to those who claim that they fail to engage with the real world of policy and practice. Such claims tend to be made on the basis of ill formed judgements, which lack evidence about what geographers are doing, or how and where geographical ideas are making a difference to policy and practice." (705)

This paper picks up the connection between sub-disciplines and epistemological traditions by exploring recent attempts to incorporate methods and theories from other (sub)disciplines into urban geography's examination of policy failure.

The paper begins by comparing the study of policy failure in urban geography and public policy. Public policy offers urban geographers a range of tools for understanding policy failure (see Bovens & 'T Hart, 2016; Dunlop, 2017; Howlett, Ramesh, & Wu, 2015; McConnell, 2010). However, in illustrating the epistemological orientations of the two fields, the paper shows how the two subdisciplines tend to develop distinct knowledges. Urban geographers are often concerned with the development of theoretical explanation (Bridge, 2014), whereas public policy is concerned with understanding governmental action within specific contexts (Gibbons, 2006; Taylor, 1989). To illustrate the consequent epistemological differences, the paper develops two contrasting interpretations of the 2008 bankruptcy of the City of Vallejo, California (Davidson & Kutz, 2015). In conclusion, the paper discusses the need for urban geographers to think about policy failure reflexively and calls for an assessment of how epistemological choices impact utility.

Epistemic communities and explanations of policy failure

Across the social sciences and humanities, there is an ongoing conversation about the potentials and pitfalls of interdisciplinary research (see Jacobs, 2014). This conversation often highlights the problems associated with bringing together research communities who operate with differing epistemologies. Some have suggested that an embrace of "epistemic pluralism" is permissible and necessary due to the representational limits of language (Lyotard, 1984) and the role of epistemological privilege in colonialism (Teffo, 2011). Others are more cautious (Boghossian, 2007). Brister (2016, p. 89) has argued that epistemological traditions create significant challenges for interdisciplinary research, highlighting "disciplinary capture" as a pressing problem. This problem involves facets of a disciplinary approach conditioning the overall design and findings of interdisciplinary research: "Disagreements about facts, evidentiary standards, the nature of causal claims, and the role of values are often exacerbated through the research process because they form integrated bundles of self-reinforcing epistemological commitments and beliefs." As urban geographers mine other fields to study policy failure, an concern about disciplinary capture is pertinent. We must identify (a) the modes of reasoning in urban geography and related fields, (b) understand how epistemological frameworks orientate us towards certain forms of explanation, and (c) develop the requisite practices of epistemological reflection.

There is no single understanding of policy failure within urban geography (see Jacobs, 2012; McCann & Ward, 2015). However, there is a predominant concern with how urban policy is determined by neoliberal ideologies. This has meant policy failure is usually discussed in the context of certain policies producing problematic social outcomes, as opposed to analysis assessing whether policy objectives are achieved (e.g. Hubbard & Lees, 2018). Where more detailed policy analyses are performed, urban geographers have tended to view policies as derivative of governmental context (Cook, 2015). This often shifts focus onto policy programs that are acutely reflective of prevailing regulatory regimes. Rarely are policies studied as practice-based interventions, where cause and effect are assessed in isolation from broader processes of social (re)production. Urban geography therefore tends to understand policy success/failure using heterodox politico-economic theories that situate policy outcomes within, and as resulting from, processes of economic and social reproduction (Brenner, Peck, & Theodore, 2010).

This reflects Bridge's (2014) view that "[T]he theoretical wellspring of critical theory in urban studies has been Marxism and neo-Marxist theory" (1–2; also see Oswin, 2018). Bridge (2014) elaborates by locating urban geography's theoretical orientation within the Western Marxist traditions of the Frankfurt School:

" ... in this approach critique had to be theoretical and separate from practical reason because practical reason operated in everyday contexts of domination and deceit. Critical theory was comprehensive and scientific and beyond the limits of lay knowledge. It did not require validation from any particular audience. This was a moment of epistemological privilege. The task, then, was to take the critique to others who may well have false beliefs about their practices." (4-5)

Frankfurt School theorists, such as Horkheimer and Adorno, famously took up the challenge of developing Marx's concepts of alienation and false consciousness at the dawn of advanced capitalism. As Bridge (2014) suggests, this meant critiquing exploitative social processes that the working classes could not themselves identify. In the absence of exploitation being an intelligible part of the everyday, Horkheimer and Adorno (2002[1947]) would work to reinvigorate theoretical explanation, claiming:

" ... the blocking of the theoretical imagination has paved the way for political delusion. Even when people have not already succumbed to such delusion, they are deprived by the mechanisms for censorship, both the external ones and those implanted within them, of the means of resisting it" (xvi).

Although Bridge (2014) identifies theoretical reasoning as a distinguishing part of the critical theory tradition (also see Brenner, 2009), the distinction between practical and theoretical reason does not originate in mid-twentieth century Western Marxism. In philosophy, the distinction is often associated with Aristotle's division of knowledge into practical and theoretical (Anagnostopoulos, 1994). Disciplines that are practically orientated focus on understanding how to act. This imposes conditions on what kind of knowledge is valued. Crucially, it demands that knowledge be specific; it being able to inform the situated complexities of practical action. Theoretical disciplines place emphasis on understanding causation and therefore tend towards producing explanations that are abstract from situated complexities (Jay, 2014). This demands rigorous, logical explanation (ibid.), but not the requirement to inform practical action. For example, when a fiscal policy fails, practical reasoning would focus on the actions that led to failure; accounts left unfiled, poor investment decisions etc. Theoretical disciplines would seek to explain why certain conditions permitted various forms of failure; poor democratic processes, neoliberal fiscal disciplining etc.

These explanatory differences are not necessarily open to synthesis since they serve different purposes; one regulating action and the other regulating belief (Jay, 2014). Any effort to synthesize away this difference confronts the problem that belief and action are not always congruent (Anagnostopoulos, 1994; Kant, 1788[2009]; Sen, 2009). Put differently, the limits to knowing demand different types of reasoning: "The distinction between theoretical and practical knowledge is based primarily on a finite manner of knowing and in terms of two basic kinds of objects: a necessary, non-operable object and a contingent, operable object." (Oesterle, 1958, p. 161). The promise of importing public policy concepts and methods into urban geography is, in large part, that they can inform and deepen our theoretical understandings of urban policy. But these imports are generated within epistemological communities who conceptualize the object of analysis differently. The result is, as the following accounts demonstrate, that more fine-grained analyses of policy failure can serve to complicate, and not enhance, our understandings of how and why urban policies fail.

Theoretical explanation in critical urban geography

Brenner's (2009, p. 201) description of critical urban theory demonstrates its theoretical orientation: "It is characterized by epistemological and philosophical reflections; the development of formal concepts, generalizations about historical trends; deductive and inductive modes of argumentation; and diverse forms of historical analysis." The two theories that have been central to studying policy successes (and failure) in urban geography have been the Marxian theory of accumulation and neoliberalism (Brenner, 2009; Bridge, 2014). David Harvey's (1978, 1989)) work on the urban process under capitalism is exemplary. Harvey's (1978) theory of capitalist urbanization begins with the following postulates: "I hang my interpretation of the urban process on the twin themes of accumulation and class struggle. The two themes are integral to each other and have to be regarded as different sides of the same coin-different windows from which to view the totality of capitalist activity" (101). This is a textbook example of theoretical reasoning. The framework conditions subsequent inquiry, making questions relating to why certain actors might perceive, for example, housing development necessary (e.g. supply/demand, affordability, equity, slum clearance etc.) ideological concerns, since drivers of the urban process have already been deduced.

Harvey's (1978) seminal theorizations have been extensively developed (see Jessop & Sum, 2000; McGuirk & Maclaren, 2001). Iterative developments of the framework have introduced new understandings of urban policy and created a sophisticated vocabulary for studying urban political transformation (Peck, 2014a). City government is now contextualized as constitutive of macro-economic and politico-ideological changes (Brenner et al., 2010; Peck, 2014b), enabling urban geographers to understand how neoliberal ideology has converted into governmental demands and incentives at the local level. Notably, this involves the assumption that cities are conditioned to be "entrepreneurial" (Harvey, 1989):

> "If, for example, urban entrepreneurialism (in the broadest sense) is embedded in a framework of zero-sum inter-urban competition for resources, jobs, and capital, then even the most resolute and avant-garde municipal socialists will find themselves, in the end, playing the capitalist game and performing as agents of discipline for the very processes they are trying to resist" (Peck, 2014a, p. 5)

With urban policy formulation and implementation derivative of neoliberal capitalism (see Lauermann, 2018), studies of policy failure in urban geography have generally taken three forms.

First, a voluminous literature now documents how economic reforms implemented over the last four decades have failed to deliver on their promises (see Harvey, 2005). At the city scale, global capitalism has been read as manifest in urban development that ignores social need and prioritizes rent extraction. For Clark (2014), urban developmental policies have valorized dubious investment over apparent social needs: " ... with the speculative construction of place rather than amelioration of conditions within a particular territory as its immediate (though by no means exclusive) political and economic goal" (8). Second, and related, urban geographers have documented the deleterious social outcomes of urban entrepreneurialism (Xue & Wu, 2015). Policies often proclaimed as "successful" have been shown to heighten social disparities, stoke social antagonisms and inflict harm on marginalized communities (Barnes, Waitt, Gill, & Gibson, 2006; Dikeç, 2006; Lees, 2008; Wyly & Hammel, 1999). Third, urban geographers have shown how policy failure has not stymied policy mobility where urban elites have found use for certain policies (McCann & Ward, 2011). Here, the role of celebrity consultants such as Richard Florida and Ed Glaeser are viewed as important in generating the ideological cover necessary to impose revanchist renewal agendas (Davidson & Iveson, 2015; Peck, 2014a, 2014b).

Practical explanations in public policy scholarship

Where many urban geographers have focused upon understanding how neoliberal capitalism has shaped urban policy (see Hackworth, 2006), public policy scholars have been interested in how failure occurs during formulation and enactment. The public policy literature has focused on questions of action and practice: how policies do, or do not, generate the desired changes. The result is a public policy literature that features growing analytical sophistication (Howlett et al., 2015), but no agreement over what causes policy failure (see McConnell, 2015). This contrasts to the critical urban geography literature, where there is significant agreement on why (neoliberal) policies fail (see Storper, 2016).

In their recent review of the public policy literature on policy failure, Howlett et al. (2015) identify the three basic conceptual frameworks used (also see: Bovens & 'T Hart, 1995; McConnell, 2010, 2015): "The earliest writing on the subject of policy failure conceived of policy success and failure either as purely technical issues amenable to easy solution [...], as highly complex politico-administrative phenomena resistant to change [...], or as purely relativistic constructions or interpretations impossible to address in any meaningful way ... " (Howlett et al., 2015). The three concepts – technical failure, institutional failure and subjective failure – all make failure a function of problematic actions within the policy process. Resources can be misallocated due to errors (Brudney & England, 1982), institutions difficult to reform due to engrained bureaucratic cultures (Brown, 2005; Dunleavy, 1995; Scharpf, 1986), and policy outcomes are hotly debated (McCann, 2002). Since the 1990s, these frameworks have been used to develop a series of further distinctions. They have identified variables to explain policy failure (Howlett et al., 2015), classify types of failure (Guy, 2015; McConnell, 2015), and separate out different dimensions of policy failure (McConnell, 2010).

The work of Allan McConnell (2010, 2015, 2016)) has been particularly instructive (see Little, 2012 for critique). In 2010, McConnell claimed "[T]he policy sciences lack an overarching heuristic framework which would allow analysts to approach the multiple outcomes of policies in ways that move beyond the often crude, binary rhetoric of success and failure." (346). In response, McConnell (2010, p. 346) proposed we divide policy "in process, program and political dimensions" and assess success/failure in each dimension (also see Howlett, 2012). The conceptualization of policy failure is therefore separated, with different stages of the policy process being assessed individually according to the following criteria: (i) the policy should achieve its goals, (ii) not be overwhelmingly criticized and (iii) gain widespread support.

Dissecting the policy process in this way has led to various types of failure being identified. While different adjectives have long been applied to failed public policies (Dunleavy, 1995; Guy, 2015), Howlett et al. (2015) recently developed codified descriptions:

"These included situations whereby good plans are not executed properly; those where good execution is wasted on poorly developed plans; those where poor planning and poor execution lead to very poor results and those where even the most rigorous analysis and execution still did not result in the achievement of goals, against all reasonable expectations, due to limitations in the existing policy paradigm."

These typologies (see Howlett et al., 2015; McConnell, 2010) demand that policy failure be precisely described. For Howlett et al. (2015) six dimensions of policy failure must be considered: extent, duration, visibility, avoidability, agreement and corruption. For each case of policy failure, Howlett et al. (2015) argue we can identify and measure the: (i) extent of the failure, (ii) how long it lasted, (iii) how much public attention it garnered, (iv) how avoidable the failure was, (v) how much agreement within a particular community exists with regards to whether the policy failed, and (vi) the amount of corruption (e.g. crime, fraud) that contributed to the failure. If empirical indicators can be developed for each dimension, the possibilities of systematically analyzing and comparing policy failures are prospectively enhanced.

This disaggregation of policy failure creates significant theoretical challenges. McConnell (2015) has claimed that although public policy analysis has developed a rich appreciation of the complexities of policy failure, it has not delivered a scientific approach. He argues that: "once we conceive of studying policy failure as 'art and craft', we are better placed to navigate the messy realpolitik of types and degrees of failure, as well as ambiguities and tensions between them." (1). Such conclusions suggest that although the public policy literature has sought to bring precision to the study of policy failure, it has produced few convincing theoretical explanations.

With this detail-orientated analysis failing to produce verifiable data and testable theories (McConnell, 2015), the public policy literature continues to find it difficult to identify what actions cause policies to fail. This presents a problem for urban geographers interested in using political science concepts and methods in their examinations of policy failure (see Lovell, 2019). The classificatory schemes of political scientists cannot be expected to add detail and nuance to urban geography frameworks since they have not delivered empirically-derived theoretical conclusions. This tension reflects how public policy scholars have operated with different epistemological assumptions compared to urban geographers. Public policy scholars have sought to identify incidents of failing action (i.e. practice) within the policy process. Policy failure is examined as an internal affair, embedded in the practical actions of government. Urban geography has considered policy failure chiefly from the perspective of critical theorizations of capitalist urbanization (Bridge, 2014). This approach does not tend to concern itself with whether a policy achieves its stated objectives, directing attention towards how policies serve, in success or failure, the structural politico-economic processes that drive policy-making (Peck, 2014a).

As urban geographers become concerned with policy failure and search out conceptual and methodological resources in other disciplines (Cook, 2015; Lovell, 2019) it is necessary to assess the extent to which these resources contain different epistemological assumptions. When we acknowledge that importing conceptual and methodological resources impacts upon the questions we can ask about policy failure, we must also face the question of what we want to know about policy failures. To illustrate this need for epistemological reflection alongside interdisciplinary conceptual and methodological borrowing, the next sections of the paper develop two contrasting interpretations of a recent policy failure: the 2008 bankruptcy of the City of Vallejo, California.[2] The two interpretations show how epistemic differences produce different research problems and forms of explanations.

The city of Vallejo's 2008 bankruptcy

The City of Vallejo has a diverse population of approximately 120,000 and is situated on the northern San Francisco Bay. Until the early 1990s, the city was home to a major shipyard and naval base. When the US Navy left town in the mid-1990s, Vallejo ran into fiscal problems. Disinvestment stimulated many attempts at renewal, but these did not replace the shipyard economy. Of course, disinvestment is not unique to Vallejo. When the city filed for bankruptcy in 2008, it reflected not just the stressors of uneven development but also the failure of the City to manage decline. Bankruptcy is one of four definitions of failure in Merriam-Webster's dictionary and Vallejo's bankruptcy has been described as a case of failure in media (Vekshin & Braun, 2010) and scholarly (Peck, 2014b) accounts. Vallejo became the first city to file for chapter 9 bankruptcy after the 2007–8 financial crisis. Amid the Great Recession, the City faced a $17m shortfall in its $80m General Fund budget (see Figure 1). In a situation where the City did not have enough funds to cover payroll expenses, the City Council voted to make an unprecedented chapter 9 filing. Never had an American city filed for chapter 9 because it could not afford its employee salaries (Trotter, 2011).

Two different interpretations of Vallejo's bankruptcy are now presented. Given the space constraints these interpretations are illustrative, not comprehensive. In the first

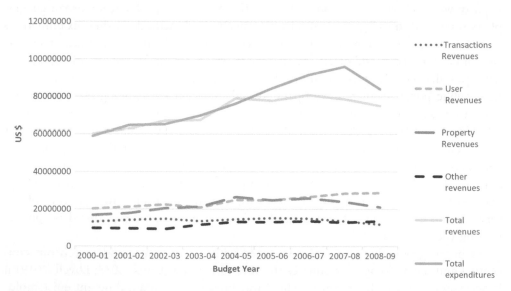

Figure 1. Breakdown of Vallejo's revenues and overall expenditures, 2000–2009.
(Source: City of Vallejo Annual Budget Statements, 2002–3-2010–11).

interpretation, the bankruptcy is presented as a consequence of speculative tendencies inherent within the entrepreneurial urban system (Davidson & Ward, 2014; Harvey, 1989; Peck, 2014a). In the second interpretation, attention focuses on policy decisions leading up to and during bankruptcy to search out specifically where failure occurred.

Take one: a failing neoliberal paradigm

Between 2005 and 2008, Vallejo's General Fund ran yearly deficits of $3.2m, $4.2m and $4.2m, leaving the City with no reserves to face a 2008 budget deficit (Mayer, 2008). In early 2008, the City's staff were projecting a $13.8m deficit in the General Fund, with larger deficits forecast for subsequent years. In the context of the Great Recession and California's property tax limiting Proposition 13, substantially increasing revenues seemed unlikely and so the City Council focused on winning concessions from labor unions (City of Vallejo, 2006). The recession left the City with two options, drastically restructure labor agreements or insolvency.

Vallejo's insolvency reflects what Davidson and Ward (2014) describe as the speculative character of contemporary entrepreneurial urban governance. They use the concept of speculative urbanism to describe the recent growth of financial risk in municipal budgets (also see Peck, 2011): "Cities have had to indulge in ever more risky forms of speculative urbanism, understood here as the ways in which cities speculate on future economic growth by borrowing against predicted future revenue streams to make this growth more likely." (ibid. 84) In Vallejo, the growth of financial risk meant that the City had consistently been betting on its ability to raise new revenues. Critically, this revenue growth was not predictable, but would come from cyclical sources given the constraints of California's governance regime (see Bardhan & Walker, 2011). Speculation was therefore based on the City's fees, permits and taxes. Two parts of this speculative budgeting are particularly important to examine here. First, growth of City revenues had become focused in predictably cyclical sources. Second, the City's economic development efforts had continually struggled to successfully utilize the tools of the neoliberal urban system.

Figure 2 shows a breakdown of Vallejo's revenues between 2000 and 2009. Revenues are divided into four categories: transaction taxes (sales and hotel taxes), user taxes (licensing and user taxes), property-related taxes (real estate taxes and property development related taxes) and other taxes (miscellaneous). Revenues are distinguished in this way to decipher taxes that are cyclical (i.e. transaction taxes), set by government assessments (i.e. user taxes), related to property markets (i.e. property-related taxes) and other smaller streams that are less impactful to the city's bottom line (i.e. miscellaneous). Between 2000 and the city's bankruptcy, property-related taxes had grown 94%, compared to 6.3% for transaction taxes, 18.9% for user taxes and 30.6% for other taxes. As the City struggled with year on year deficits, its financial wellbeing had therefore become reliant on a buoyant property market and new housing construction.

Record transfer taxes and permit fees enabled the City to meet its spending commitments. In 2007, Vallejo's property-related and transaction tax revenues started to decline and, consequently, the city's deficit became an intractable problem. In 2004–5 and 2005–6, at the peak of the housing bubble, Vallejo collected over $5m in Property Transfer Taxes (PTT). In 2007–8, PTT revenues were below $1.7m. Likewise,

Type	Budget Year	2000-01	2001-02	2002-03	2003-04	2004-05	2005-06	2006-07	2007-08	2008-09
Transaction	Sales Tax	11,304,600	12,056,800	12,742,000	12,145,303	13,156,015	13,819,405	13,353,505	12,021,086	10,467,821
	Transient Occupancy Tax	1,928,200	2,125,300	2,100,000	1,447,810	1,402,835	1,405,410	1,618,954	1,497,237	1,328,873
	Transactions Revenues	13,232,800	14,182,100	14,842,000	13,593,113	14,558,850	15,224,815	14,972,459	13,518,323	11,796,694
	Year on Year Change	-	7.2%	4.7%	-8.4%	7.1%	4.6%	-1.7%	-9.7%	-12.7%
Government Assessment	Utility Users Taxes	11,109,600	11,209,600	12,000,000	11,707,589	11,749,465	12,488,855	12,504,321	13,208,564	12,766,945
	Franchise Taxes	1,912,300	2,169,900	2,248,000	2,289,454	2,344,994	2,377,793	3,061,529	3,992,171	4,866,294
	Business License Tax	1,014,100	1,229,000	1,337,000	1,218,595	1,323,987	1,298,046	1,388,111	1,364,571	1,533,454
	Motor Vehicle License Fees	6,193,400	6,596,400	6,843,800	5,688,734	9,523,694	8,592,520	9,536,759	9,850,561	9,492,807
	User Revenues	20,229,400	21,204,900	22,428,800	20,904,372	24,942,140	24,757,214	26,490,720	28,415,867	28,659,500
	Year on Year Change	-	4.8%	5.8%	-6.8%	19.3%	-0.7%	7.0%	7.3%	0.9%
Property	Property Tax	10,039,700	11,243,200	11,761,900	12,681,006	13,623,535	15,857,808	18,776,182	19,473,533	17,670,610
	Property Transfer Tax	2,470,100	2,379,800	3,214,800	4,020,000	5,481,108	5,106,488	3,778,090	1,696,396	1,973,068
	Real Property Exercise Tax	998,500	1,142,000	1,859,900	842,000	2,054,766	256,438	662,491	91,039	44,770
	Development Fees and Permits	3,280,200	3,023,400	3,589,600	3,787,919	5,300,475	3,543,898	2,578,731	2,613,218	1,329,205
	Property Revenues	16,788,500	17,788,400	20,426,200	21,330,925	26,459,884	24,764,632	25,795,494	23,874,186	21,017,653
	Year on Year Change	-	6.0%	14.8%	4.4%	24.0%	-6.4%	4.2%	-7.4%	-12.0%
Other	Other revenues	9,850,000	9,696,400	9,354,300	11,691,590	13,173,488	13,076,709	13,578,086	12,866,059	13,576,336
			-1.6%	-3.5%	25.0%	12.7%	-0.7%	3.8%	-5.2%	5.5%
	Total revenues	60,100,700	62,871,800	67,051,300	67,520,000	79,134,362	77,823,370	80,836,759	78,674,435	75,050,183
	Total expenditures	58,898,800	64,733,200	65,191,784	69,873,353	76,308,950	84,467,987	91,579,625	96,026,974	84,003,809

Figure 2. City of Vallejo's audited revenue streams, 2000–1 thru 2008–9.
(Source: City of Vallejo annual budget statements, 2002–3–2010–11).

Development Fee and Permit revenues peaked in 2004–5, with over $5m being collected. In 2008–9, this revenue fell below $1.4m. Deficits that could be managed by drawing down reserves in previous years (Mayer, 2008) had become large enough to cause insolvency. Reliance on property-related revenue growth proved unsustainable. In this respect, the City's bankruptcy appears a consequence of the speculative governance paradigm: when predicted future revenues did not materialize (see Davidson & Ward, 2014) it became a clear candidate for neoliberal austerity restructuring (see Peck, 2014a).

The second dimension of the speculative paradigm relates to Vallejo's economic development program. The most significant economic event in Vallejo's recent past was the closure of the Mare Island Naval Base in 1996. Shuttering the 5,200-acre military facility meant Vallejo lost its largest single employer and had to begin significant economic restructuring. Many policymakers, bureaucrats and citizens associate the City's bankruptcy with its inability to establish an economic future after the shipyard closure. As a City Councilor described:

"This is a long-term turnaround that will take at least 20 [more] years to achieve. It involves changing the culture of the city. When I first moved here, people were quite happy being a military town. They really didn't care about what was happening outside. It was quite a cut off place ... Many people still have the mindset that we will rebuild industry on Mare Island, that the gas facility or a shipbuilder is coming in. It is hard to get people to think differently about development, having us be more creative ... " (Interview T3, 2017)

The City has been successful in supporting a university campus, light industry and new housing development on Mare Island, but most of the peninsula awaits redevelopment. Vallejo is just one city among many that continue to struggle with post-industrial economic restructuring.

This lack of redevelopment might extend to policy failure when an assessment of the City's use of development vehicles is considered. Unlike many other Californian cities that used Regional Development Agencies (RDAs) to channel development dollars into the city budget, Vallejo's RDA has been a minor actor in development efforts (Davidson & Ward, 2014). Indeed, in bankruptcy court documents (Mialocq, 2008) it was noted that the City's General Fund had continually subsidized the economic development efforts of the RDA. This contrasts to many Californian RDAs that subsidized the general operations of city government (see DeHaven, 2017). Vallejo has also been a relatively conservative user of TIFs (Davidson & Ward, 2014). Again, in contrast to many cities where TIFs have generated significant, if somewhat contentious, economic development. Vallejo therefore did not use all the development mechanisms afforded by the US urban system. Those who have worked with the City explain this failure as being caused by a lack of bureaucratic support, poor long-term planning and an inability to secure large development deals.

This whistle-stop account of the entrepreneurial characteristics of Vallejo helps to demonstrate how the City's fiscal failure can be related to its governance methods. The City had continually relied on speculative revenues to produce balanced budgets, become dependent on cyclical revenue growth, and not been successful in using neoliberal urban development tools. After bankruptcy, the prevailing entrepreneurial system provided little scope for Vallejo to transform its budget bar austerity and/or more speculation (see Peck, 2014b). So, can we call this a policy failure? If yes, precisely what policies failed, and how? It is at this point that the frameworks developed in public policy might offer help. However, the addition of theoretical and methodological tools from public policy brings with it the problems of epistemological difference.

Take two: the process, programs and politics of policy failure

McConnell's (2015) three-part characterization of policy failure – process, program and politics – can be used to study Vallejo's bankruptcy. An evaluation of "process" is concerned with how governments get approval for policies and develop standards of assessment; in the case study, how bankruptcy became the policy. "Program" evaluation relates to those specific policies that are "designed to address goals and underpinned by assumptions about appropriate levels of government intervention in society" (ibid.). For example, how the City of Vallejo used bankruptcy to correct its fiscal problems. Finally, "political" is evaluated by the ways in which political conflicts are managed by govern-ments to generate the desired outcomes; for Vallejo, if support for government policies was generated.

Process

Before the Great Recession, the City of Vallejo was already experiencing budget stress. On the revenue-side, the City saw a constrained environment. The City claimed: "While we do exercise a level of control over our expenses, in some cases we have little or no control over our revenues" (City of Vallejo, 2006: iv). When looking to raise revenues, the City argued: "[W]hile we are examining a variety of potential new fees and current fee increases, our options for raising revenue are limited by recent court rulings and the need for voter-approved increases (either a simple majority or a two-thirds vote, depending on the tax). Our only realistic option for balancing the budget is to reduce expenditures" (ibid. vi). With the first signs of the housing bubble bursting, the City was already concerned that revenue growth was slowing.

In the 2006–7 City Budget, around 90 percent of the General Fund had been allocated to employee salaries and benefits. In preceding years, the City had cut staffing levels and reduced operating funds to work within personnel budgets, but this was now not enough. The City therefore presented itself with a stark choice: "The discussion of a reduction in expenditures is a relatively simple one as well, as we really have two choices: cut the budgets of various categories (which, in essence, means a reduction in staffing, as that is our biggest cost) or work to obtain cooperation from employee groups on cost-cutting ideas" (ibid. vi). Although some city councilors and citizens had long voiced concerns about the power of labor unions and their generous bargaining agreements (City of Vallejo, 1993), slowing revenue growth was impressing on the City the need to make labor reforms before the Great Recession.

In early 2006, the City Council motioned to start building a 15 percent revenue reserve. The 2006–7 City Budget stated the intent to make labor agreement reforms. In 2007 and 2008, the City undertook contentious negotiations with its labor unions. This process took various forms, including mediation and arbitration hearings (see McManus, 2008). In terms of evaluating the "process" of fiscal reform leading up to bankruptcy, actors on both sides have different interpretations of its effectiveness. Labor union representatives claimed that the process worked, pointing towards the reforms offered by labor groups that would, according to them, have balanced the budget (ibid.). The fault, for labor representatives, therefore rested with the City Council and its desire to side-step conventional bargaining processes. On the City Council side, the process was broken. Specifically, a chartered commitment to binding arbitration was seen to place the City in an impossible position. As one councilor claimed: "Whatever happened in those negotiations, we always had our hands tied with binding arbitration ... The City had never won a binding arbitration case, so we knew that we had little leverage until we removed the commitment from the [City] Charter." (Interview C2 2017). On 6 May 2008, Vallejo's City Council voted unanimously to file for chapter 9 bankruptcy.

Program

Fiscal reform using chapter 9 therefore became the policy of the City of Vallejo. Given Vallejo's largest fiscal liability was labor-related expenditures, the City had to win the court's approval that minimum staffing levels and collective bargaining agreements (CBAs) could be subject to chapter 9 restructuring. This was a complex and contentious issue (see McManus, 2008). However, Vallejo was granted the right to change CBAs in its bankruptcy readjustment plan. If the focus of the bankruptcy was the reduction of labor-related expenditures, it is therefore necessary to assess whether reductions occurred and if the reforms created balanced budgets.

The City opted to renegotiate some CBAs outside of bankruptcy. In 2008, the City came to new CBA agreements with the Vallejo Police Officers' Association (VPOA) and the Confidential, Administrative, Managerial and Professional (CAMP) labor unions. It would later come to an agreement with the International Associations of Firefighters union (IAFF) and, in the bankruptcy settlement, impose terms on the International Brotherhood of Electrical Workers (IBEW). The result was an overall reduction in labor expenditures but an uneven treatment of workers. Figure 3 shows that VPOA and CAMP employees received salary increases in 2010–11, with only IBEW members seeing salary cuts. New hires on IAFF and IBEW contracts would receive fewer benefits. The most significant reductions came from across the board cuts in retiree health benefits, with existing variable rates being replaced by a capped ($300/mo) benefit. CBA reductions were also paired with continued cuts in staffing levels. From a pre-bankruptcy high of 155 police officers, in 2010 there were 92 sworn officers in Vallejo. The City also reduced its fire companies from nine to five.

Average cost per employee	Police Union - VPOA	Fire Union - IAFF	Electrical Workers Union - IBEW	Administrators Union - CAMP (Mid-Managers)	Executive (Dept Directors)
Salary, including various differential pays	122,546	115,708	65,858	105,925	185,016
CalPERS pension (normal cost and UAAL)	39,983	32,531	10,990	19,944	43,609
Health/Welfare Benefits	16,642	13,505	12,457	14,178	10,218
Retire Health (normal cost and UAAL)	17,170	7,322	3,580	13,353	7,215
Workers Compensation	21,445	17,264	2,892	2,598	10,623
Other	1,694	1,773	1,802	7,522	7,838
	219,480	188,103	97,579	163,520	264,519
Salary - COLA					
Salary Increase (decrease)	7%	0%	-5%	2%	0%
Water Treatment and Communication Operators			10%		
Furlough Days				6 days (2.3%)	
Pension Benefits					
Existing Employees	3% @ 50	3% @ 50	2.7% @ 55	2.7% 55	2.7% 55
2nd Tier for New Hires		2% @ 50	2.0% @ 55		
Contribution Rate - City	32.66%	28.26%	17.02%	18.82%	18.82%
Contribution Rate - Employee	9.00%	13.40%	10.80%	9.00%	9.00%
	41.66%	41.66%	27.82%	27.82%	27.82%
Health/Welfare Benefits					
Medical (Share of Kaiser rate, including Cafeteria Plan)	100%	75%	75%	100%	75%
Average cost per employee (varies with dependents)	$14,306	$11,025	$9,733	$11,512	$8,034
Vision/Dental	100%	100%	100%	100%	75%
Average cost per employee (varies with dependents)	$2,235	$2,235	$2,038	$1,994	$1,526
Other - Life, AAD, and/or LTD (varies by group)	$101	$245	$686	$672	$658
	$16,642	$13,505	$12,457	$14,178	$10,218
Retiree Health Benefits					
Current Benefit	100%	$300/mo	75%	80%	$300/mo
If retired before July 2000 (before 3%@50 pension)		75%			
OPEB Funding (assumes future VPOA/Camp reductions)	$300/mo	$300/mo	$300/mo	$300/mo	$300/mo
If retired before July 2000 (before 3%@50 pension)	75%	75%			
Contribution Rate (% of payroll)					
Normal Cost	1.70%	1.50%	2.40%	1.60%	1.20%
Amortization of Unfunded Liability (Includes current pay-as-go for VPOA/CAMP)	12.00%	4.90%	3.20%	11.00%	2.70%
Workers Compensation	14.10%	6.40%	5.60%	12.60%	3.90%
Self-insurance rates	17.50%	15.00%	2.4% - 8.7%	2.40%	2.40%

Figure 3. City of Vallejo, Salary and benefit assumptions, 2010–11 proposed Budget.
(Source: City of Vallejo annual budget, 2010–11).

The reforms enabled the City to substantially reduce expenditures in 2010. Although labor contract changes were not the only fiscal reform undertaken in Vallejo (see City of Vallejo, 2010), they represented the most sizable. Expenditures on salaries and benefits moved from a projected $71m to $59m. Cuts to services and supply expenditures amounted to $7m, but most of these related to deferred purchases or payments and would have to be covered in future years (Interview T4, 2013). Overall, the City's 2010–11 budget aimed to transform a 22% deficit into a 2% surplus. The audited 2010–11 budget shows that the City was able to achieve this goal. The 2010–11 fiscal year returned a General Fund surplus of $2.3m and reserves were increased to 9% of expenditures ($6.3m).

Although many budgetary changes have followed since 2010–11, bankruptcy, either directly (i.e. retiree health care benefit reform) or indirectly (i.e. CBA renegotiation), returned a degree of fiscal stability to the City. In 2013–14, the City was able to project a structurally balanced budget. However, budget forecasts relied on a continued freezing of salaries and benefits, and new pension payments rates imposed by the state pension agency will demand new revenue growth (City of Vallejo, 2017). Although Vallejo can claim limited success in its bankruptcy, the City continues to face severe fiscal problems (Raskin-Zrihen, 2017.).

Political

McConnell (2015) suggests that a successful policy will have near universal support. If this is the case, Vallejo's bankruptcy is far from an unqualified success. Although the City Council voted unanimously for chapter 9, one City Councilor claimed that: "It was only when the City Council were told that they might become personally liable for the city's debts that support for the filing became unanimous" (Interview T6 2017). Although there was no public vote to approve bankruptcy, the City has recently undertaken two popular referendum on fiscal matters. These give some indication of how much public support there has been for the City's fiscal reforms.

The first referendum related to the article in the City's Charter that committed the City to binding arbitration in CBA conflicts. In June 2010, residents of Vallejo voted on the following question:

> "Shall Section 809 of the Charter of the City of Vallejo be repealed to remove the mediation/arbitration process, commonly referred to as binding interest arbitration, that permits an arbitrator, without City Council approval, to make the final decision to resolve disputes between the City and its recognized employee organizations on all matters relating to wages, hours and working conditions and instead to use the method of resolving such disputes set forth in state law"

Only 24 of the 478 Californian cities have City Charter commitments to binding arbitration. It was thought by some members of the City Council that Vallejo's commitment to binding arbitration had been a key factor in the bankruptcy. The City Council voted 6–1 to place the measure on the ballot. The measure received 9,314 (51.12%) "Yes" votes and 8,856 (48.74%) "No" votes. By a narrow majority, the City of Vallejo was able to remove binding arbitration from its Charter.

The second referendum related to a new sales tax proposed by the City Council. Placed on the November 2011 ballot, residents were asked to vote on the following:

"To enhance funding for 9-1-1 response, police patrols, firefighter and paramedic services, youth and senior programs, street and pothole repair, graffiti removal, economic development, and general city services, shall the sales tax be raised one cent, expiring after ten years, with all revenue legally required to stay in Vallejo?"

The measure would raise the sales tax in Vallejo from 7.375% to 8.375%. The additional tax was to be limited to 10 years and revenue allocated to services that had degraded in previous years. The latter was thought particularly crucial by advocates (Interviews 2013, 2017) so residents would trust that tax dollars would reach areas of need. The Measure B ballot returned 9,295 (50.43%) "Yes" votes and 9,136 (49.57%) "No" votes. The City has subsequently expanded public services using the new sales tax monies and made the measure permanent.

In terms of gauging the political success of the City's fiscal programs, the two referenda demonstrate a consistent split within the city. Although these referenda offer no direct indication of political support for bankruptcy reforms, they do show how related fiscal reforms have relied on a small majority to move forward.

Using McConnell's (2015) three-part evaluation scheme for policy failure delivers mixed answers. The process of arriving at bankruptcy as a policy program was decidedly messy, appearing a consequence of entrenched local politics as much as entrepreneurial governance. Contentious labor negotiations preceded an economic downturn that transformed a pressing problem into something requiring triage. As bankruptcy became the "program" emerging from the "process", it ultimately delivered some, if not complete, fiscal stability. Finally, indicators of political support for the City's fiscal reforms present a mixed picture. Our diagnosis of failure is messy and does not illustrate an obvious connection between policies and the governance regime.

Conclusions

A concern with policy failure in urban geography has led to interdisciplinary experimentation (Cook, 2015; Lovell, 2019) and consequently generated epistemological challenges. As the contrasting interpretations of the Vallejo bankruptcy show, distinct epistemological traditions construct different inquiries into and explanations of urban policy failure. In the case of urban geography, a long engagement with critical social theory (Brenner, 2009; Bridge, 2014) has given the subdiscipline's concern with urban policy a theoretical orientation (Imrie, 2004). Producing knowledge to immediately inform the (practical) actions of city governments has not therefore been an overriding priority (ibid.). This contrasts to the practically-orientated public policy literature, where attention has focused on knowing where government action goes wrong and how government action might be improved (Howlett et al., 2015).

These differences demonstrate the contrasting objectives of practical and theoretical reasoning (Anagnostopoulos, 1994; Oesterle, 1958). Practically-orientated inquiry focuses on the regulation of action. Attempts to generate knowledge to improve action tend to lack a concern with causation, since the intent is to produce concrete knowledge with relevance to practice. Theoretical inquiry differs in that a concern with causation can mean the production of abstract knowledge that is abstruse for the purposes of acting. For example, if we supplement the geographical explanation (i.e. speculative entrepreneurialism) of Vallejo's policy failure with an approach derived from public policy, we likely move from a rather unambiguous explanation of causation to a more complicated story of fragmented policy formulation and implementation. Explaining policy failure as a consequence of neoliberalism and informing practical action are often incompatible. It is not that either form of knowledge is invalid (ibid.), but rather that they are not necessarily open to synthesis. This acknowledgment of epistemic pluralism not only problematizes the un-reflexive combining of different epistemological traditions (Brister, 2016), but also signals the need for debate about (a) why we are now concerned with policy failures, and (b) what it is we want to know about policy failures? Only by answering these questions can the study of policy failure navigate epistemological difference and make informed choices about epistemological orientation.

Such reflections can lead down many different paths. One such path might be epistemological anarchism (see Feyerabend, 1975). However, an acceptance of epistemological pluralism can be negotiated without resort to relativism. Pragmatist philosophy can be instructive in this regard (also see Marchart, 2007 on post-foundationalism). Pragmatists have long argued that truth is closely related with utility. Rorty (1992, p. 582) argued that rational, scholarly inquiry involves the application of technical reason for the enhancement of tolerance and, thus, freedom. The search for validity in different epistemological perspectives therefore " ... only looks relativistic if one thinks that the lack of general, neutral, antecedently formulable criteria for choosing between alternative, equally coherent, webs for belief means that there can be not 'rational' decision. Relativism seems a threat only to those who insist on quick fixes and knock-down arguments." (Rorty, 1991, p. 66). Rorty goes onto argue that we do not have a duty to formulate general epistemological principles, rather we have "a duty to talk to each other, to converse about our views of the world, to use persuasion rather than force, to be tolerant of diversity, to be contritely fallibilist" (ibid. 67).

When faced with a choice of epistemological approach, Rorty suggests we can only make decisions over which to work within by "running back and forth between principles and the results of applying principles" (ibid. 68). In other words, when we change the means of our inquiries (e.g. adopt the concepts and methods of public policy to investigate policy failure) we must assess how this shift changes the ends of our inquiries. In reflecting on how new means offer different scholarly ends, you can reflexively come to know what you want inquiry to achieve: "you only know what you want after you've seen the results of your attempts to get what you once thought you wanted" (ibid. 68). Although urban geography's current concern with policy failure is the cause of significant epistemological challenges, it also offers opportunity for reflection and the forms of skepticism that have brought about modern philosophical and theoretical reorientations (see Lilla, 1993). What precisely geographers have to say about "policy failure" should not be conditioned by unquestioned epistemic traditions or entrenched views on "policy relevance" (see Imrie, 2004). The problem of "policy failure" presses urban geographers to explicitly consider the intended utilities of their inquiries. By acknowledging epistemological differences are not always open to synthesis, a consideration of the objectives and implications of inquiry must play a more significant part in emerging geographical discussions of policy failure.

Notes

1. The adage has an unclear origin. It has been attributed to the likes of Yogi Berra, Albert Einstein, and Richard Feynman.
2. The paper draws on research conducted between 2010 and 2017 that examined the various fiscal-related reforms undertaken in Vallejo. The research included the collection and analysis of city budget documents, bankruptcy filings, and secondary literature appertaining to the City's bankruptcy. Three field visits to Vallejo (2011, 2013 and 2017) were also undertaken to interview 35 key-informants (e.g. City Councilors, administrators, civic society actors, community organizers) and 11 current and past residents on issues relating to the City's bankruptcy and restructuring.

Disclosure statement

No potential conflict of interest was reported by the author.

References

Anagnostopoulos, Georgios. (1994). *Aristotle on the goals and exactness of ethics.* Berkeley, CA: University of California Press.

Bardhan, Ashok, & Walker, Richard. (2011). California shrugged: Fountainhead of the great recession. *Cambridge Journal of Regions, Economy and Society, 4*(3), 303–322.

Barnes, Kendall, Waitt, Gordon, Gill, Nicholas, & Gibson, Chris. (2006). Community and Nostalgia in urban revitalisation: A critique of urban village and creative class strategies as remedies for social 'problems'. *Australian Geographer, 37*(3), 335–354.

Boghossian, Paul. (2007). *Fear of knowledge: Against relativism and constructivism.* Oxford: Oxford University Press.

Bovens, Mark, & 'T Hart, Paul. (1995). Frame multiplicity and policy fiascoes: Limits to explanation. *Knowledge and Policy, 8*(4), 61–82.

Bovens, Mark, & 'T Hart, Paul. (2016). Revisiting the study of policy failures. *Journal of European Public Policy, 23*(5), 653–666.

Brenner, Neil. (2009). What is critical urban theory? *City, 13*(2–3), 198–207.

Brenner, Neil, Peck, Jamie, & Theodore, Nik. (2010). Variegated neoliberalization: Geographies, modalities, pathways. *Global Networks, 10*(2), 182–222.

Bridge, Gary. (2014). On marxism, pragmatism and critical urban studies. *International Journal of Urban and Regional Research, 38*(5), 1644–1659.

Brister, Evelyn. (2016). Disciplinary capture and epistemological obstacles to interdisciplinary research: Lessons from central African conservation disputes. *Studies in History and Philosophy of Biological and Biomedical Sciences, 56*(April), 82–91.

Brown, RR. (2005). Impediments to integrated urban stormwater management: The need for institutional reform. *Environmental Management, 36*(3), 455–468.

Brudney, Jeffrey, & England, Robert. (1982). Urban policy making and subjective service evaluations: Are they compatible? *Public Administration Review, 42*(2), 127–135.

City of Vallejo. (May 11, 1993). *Report of the citizens budget advisory committee.* Vallejo, CA: City of Vallejo.

City of Vallejo. (2006). *2006-7 annual budget.* Vallejo, CA: City of Vallejo.

City of Vallejo. (2010). *2010-11 annual budget.* Vallejo, CA: City of Vallejo.

City of Vallejo. (2017). *2017-18 annual budget.* Vallejo, CA: City of Vallejo.

Clark, Eric. (2014). Good urban governance: Making rent gap theory not true. *Geografiska Annaler: Series B, Human Geography, 96*(4), 392–395.

Clarke, Nick. (2012). Urban policy mobility, anti-politics, and histories of the transnational municipal movement. *Progress in Human Geography, 36*(1), 25–43.

Cook, Ian. (2015). Policy mobilities and interdisciplinary engagement. *International Journal of Urban and Regional Research, 39*(4), 835–837.

Davidson, Mark, & Iveson, Kurt. (2015). Recovering the politics of the city: From the 'post-political city' to a 'method of equality' for critical urban geography. *Progress in Human Geography, 39*(5), 543–559.

Davidson, Mark, & Kutz, William. (2015). Grassroots austerity: Municipal bankruptcy from below in Vallejo, California. *Environment and Planning A, 47*(7), 1440–1459.

Davidson, Mark, & Ward, Kevin. (2014). 'Picking up the pieces': Austerity urbanism, California and fiscal crisis. *Cambridge Journal of Regions, Economy and Society, 7*(1), 81–97.

DeHaven, James. (2017, October 13). Redevelopment dollars are ending up in the city general fund. Retrieved from http://www.sandiegouniontribune.com/news/watchdog/sd-me-housing-funds-20160111-story.html

Dikeç, Mustafa. (2006). Two decades of French urban policy: From social development of neighbourhoods to the Republican Penal state. *Antipode, 38*(1), 59–81.

Dorling, Danny, & Shaw, Mary. (2002). Geographies of the agenda: Public policy, the discipline and its (re)"turns.". *Progress in Human Geography, 26*(5), 629–646.

Dunleavy, Patrick. (1995). Policy disasters: Explaining the UK's record. *Public Policy and Administration, 10*(2), 52–70.

Dunlop, Claire. (2017). Policy learning and policy failure: Definitions, dimensions and intersections. *Policy & Politics, 45*(1), 3–18.

Feyerabend, Paul. (1975). *Against method: Outline of an anarchistic theory of knowledge.* New York: New Left Books.

Gibbons, Michael. (2006). Hermeneutics, political inquiry, and practical reason: An evolving challenge to political science. *The American Political Science Review, 100*(4), 563–571.

Guy, Peters, B. (2015). State failure, governance failure and policy failure: Exploring the linkages. *Public Policy and Administration, 30*(3–4), 261–276.

Hackworth, Jason. (2006). *The neoliberal city: Governance, ideology, and development in American urbanism.* Ithaca, NY: Cornell University Press.

Hamnett, Chris. (2003). Contemporary human geography: Fiddling while Rome burns? *Geoforum, 34*(2), 1–3.

Harvey, David. (1978). The urban process under capitalism: A framework for analysis. *International Journal of Urban and Regional Research, 2*(1–3), 101–131.

Harvey, David. (1989). From managerialism to entrepreneurialism: The transformation in urban governance in late capitalism. *Geografiska Annaler. Series B, Human Geography, 71*(1), 3–17.

Harvey, David. (2005). *A brief history of neoliberalism.* Oxford: Oxford University Press.

Horkheimer, Max, & Adorno, Theodore. (2002[1947]). *Dialectic of enlightenment: Philosophical fragments.* Stanford, CA: Stanford University Press.

Howlett, Michael. (2012). The lessons of failure: Learning and blame avoidance in public policy-making. *International Political Science Review, 33*(5), 539–555.

Howlett, Michael, Ramesh, Michael, & Wu, Xun. (2015). Understanding the persistence of policy failures: The role of politics, governance and uncertainty. *Public Policy and Administration, 30*(3–4), 209–220.

Hubbard, Phil, & Lees, Loretta. (2018). The right to community? *City, 22*(1), 8–25.

Imrie, Rob. (2004). Urban geography, relevance, and resistance to the "Policy turn". *Urban Geography, 25*(8), 697–708.

Jacobs, Jane. (2012). Urban geographies I: Still thinking cities relationally. *Progress in Human Geography, 36*(3), 412–422.

Jacobs, Jerry. (2014). *In defense of disciplines: Interdisciplinarity and specialization in the research university.* Chicago: University of Chicago Press.

Jay, Wallace, R. (2014). Practical reason. *Stanford Encyclopedia of Philosophy.* Retrieved from https://plato.stanford.edu/entries/practical-reason/

Jessop, Bob, & Sum, Ngai-Ling. (2000). An entrepreneurial city in action: Hong Kong's emerging strategies in and for (Inter)urban competition. *Urban Studies, 37*(12), 2287–2313.

Kant, Immanuel. (1788[2009]). *Critique of practical reason*. London: Seven Treasures Publications.

Lauermann, John. (2018). Municipal statecraft: Revisiting the geographies of the entrepreneurial city. *Progress in Human Geography*, 42(2), 205–222.

Lees, Loretta. (2008). Gentrification and social mixing: Towards an inclusive urban renaissance? *Urban Studies*, 45(12), 2449–2470.

Lilla, Mark. (1993). *G.B. Vico: The making of an anti-modern*. Cambridge, MA: Harvard University Press.

Little, Adrian. (2012). Political action, error and failure: The epistemological limits of complexity. *Political Studies*, 60(1), 3–19.

Lovell, Heather. (2016). The role of international policy transfer within the multiple streams approach: The case of smart electricity metering in Australia. *Public Administration*, 94(3), 754–768.

Lovell, Heather. (2017b). Are policy failures mobile? An investigation of the advanced metering infrastructure program in the state of Victoria, Australia. *Environment and Planning A*, 49(2), 314–331.

Lovell, Heather. (2019). Policy mobilities failure. *Progress in Human Geography*, 43(1), 46–63.

Lyotard, Jean-François. (1984). *The postmodern condition: A report on knowledge*. Minneapolis: University of Minnesota Press.

Marchart, Oliver. (2007). *Post-foundational political thought*. Edinburgh: University Edinburgh Press.

Markusen, Ann. (1999). Fuzzy concepts, scanty evidence, policy distance: The case for rigour and policy relevance in critical regional studies. *Regional Studies*, 33(6–7), 869–884.

Martin, Ron. (2001). Geography and public policy: The case of the missing agenda. *Progress in Human Geography*, 25(2), 189–210.

Mayer, Susan. (2008, May 23). *Declaration of Susan Mayer in support of the statement of qualifications under Section 109(c) Case No. 2008-26813 in re. City of Vallejo*. United States Bankruptcy Court, Eastern District of California.

McCann, Eugene. (2002). The cultural politics of local economic development: Meaning-making, place-making, and the urban policy process. *Geoforum*, 33(3), 385–398.

McCann, Eugene, & Ward, Kevin. (2011). *Mobile urbanism: Cities and policymaking in the global age*. Minneapolis: University of Minnesota Press.

McCann, Eugene, & Ward, Kevin. (2015). Thinking through dualisms in urban policy mobilities. *International Journal of Urban and Regional Research*, 39(4), 828–830.

McConnell, Alan. (2010). Policy success, policy failure and grey areas in-between. *Journal of Public Policy*, 30(3), 345–362.

McConnell, Alan. (2015). What is policy failure? A primer to help navigate the maze. *Public Policy and Administration*, 30(3–4), 221–242.

McConnell, Alan. (2016). A public policy approach to understanding the nature and causes of foreign policy failure. *Journal of European Public Policy*, 23(5), 667–684.

McGuirk, Pauline, & Maclaren, Andrew. (2001). Changing approaches to urban planning in an 'Entrepreneurial City': The case of Dublin. *European Planning Studies*, 9(4), 437–457.

McManus, Michael. (2008, September 5). *Findings of fact and conclusions of law. Case No. 2008-26813 in re. City of Vallejo*: United States Bankruptcy Court, Eastern District of California.

Mialocq, Roger. (2008, June 27). Declaration of Roger Mailocq in support of IAFF Local 1186, VPOA, and IBEW local 2376 ("unions") objection to Debtor's bankruptcy petition and statement of qualifications under Section 109(c) Case No. 2008-26813 in re. City of Vallejo: United States Bankruptcy Court, Eastern District of California.

Newman, Kathe, & Ashton, Phil. (2004). Neoliberal urban policy and new paths of neighborhood change in the American inner city. *Environment and Planning A*, 36(7), 1151–1172.

Oesterle, John. (1958). Theoretical and practical knowledge. *The Thomist: A Speculative Quarterly Review*, 21(2), 146–161.

Oswin, Natalie. (2018). Planetary urbanization: A view from outside. *Environment and Planning D: Society and Space*, 36(3), 540–546.

Peck, Jamie. (2011). Zombie neoliberalism and the Ambidextrous state. *Theoretical Criminology*, *14*(1), 104–110.

Peck, Jamie. (2014a). Entrepreneurial urbanism: Between uncommon sense and dull compulsion. *Geografiska Annaler: Series B, Human Geography*, *96*(4), 396–401.

Peck, Jamie. (2014b). Pushing austerity: State failure, municipal bankruptcy and the crises of fiscal federalism in the USA. *Cambridge Journal of Regions, Economy and Society*, *7*(1), 17–44.

Raskin-Zrihen, Rachel. (2017, October 13). Vallejo is nation's second least recession-recovered city, new study finds. *Vallejo Times-Herald*. Retrieved from http://www.timesheraldonline.com/article/NH/20170112/NEWS/170119905

Rorty, Richard. (1991). *Objectivity, relativism, and truth: Philosophical papers*. Cambridge: Cambridge University Press.

Rorty, Richard. (1992). A pragmatist view of rationality and cultural difference. *Philosophy East and West*, *42*(4), 581–596.

Scharpf, Fritz. (1986). Policy failure and institutional reform: Why should form follow function? *International Social Science Journal*, *38*(108), 179–189.

Sen, Amartya. (2009). *The idea of justice*. Cambridge, MA: Belknap Press.

Storper, Michael. (2016). The neo-liberal city as idea and reality. *Territory, Politics, Governance, 4* (2), 241–263.

Taylor, Charles. (1989, August). Explanation and practical reason. *WIDER working papers*.

Teffo, Lesbia. (2011). Epistemic pluralism for knowledge transformation. *International Journal of African Renaissance Studies*, *6*(1), 24–34.

Trotter, Richard. (2011). Running on empty: Municipal insolvency and rejection of collective bargaining agreements in chapter 9 Bankruptcy. *Southern Illinois University Law Journal, 36* (1), 46–87.

Vekshin, Alison, & Braun, Martin. (2010). Vallejo's bankruptcy 'failure' scares cities into cutting costs. *Bloomberg, 14*(December).

Webber, Sophie. (2015). Mobile adaptation and sticky experiments: Circulating best practices and lessons learned in climate change adaptation. *Geographical Research*, *53*(1), 26–38.

Wyly, Elvin, & Hammel, Daniel. (1999). Islands of decay in seas of renewal: Housing policy and the resurgence of gentrification. *Housing Policy Debate*, *10*(4), 711–771.

Xue, Desheng, & Wu, Fulong. (2015). Failing entrepreneurial governance: From economic crisis to fiscal crisis in the city of Dongguan, China. *Cities, 43*, 10–17.

Policy-failing: a repealed right to shelter

Katie J. Wells

ABSTRACT

In 1984 voters in Washington, D.C.approved an initiative to guarantee overnight shelter. The city was one of the first jurisdictions in the world to legislate aright to shelter and, to this day, the only one to do so by popular initiative. But the law, which emerged during one of the city's most politically progressive periods, was overturned by the local legislature, and a voter referendum failed to reinstate it. Based on archival research and 25 in-depth interviews, this paper examines how a homeless shelter policy was made to fail. This paper analyzes the work of activists, service providers, city officials, and legal actors who created, contested, and eventually repealed the law. This analysis contributes to understandings of urban governance with theoretical explorations of the notion of policy-failing and empirical attention to the institutional mechanisms, political ideologies, and regulatory practices through which policies can be un made.

Introduction

In November 1984, as voters across the USA went to the polls to re-elect neoconservative President Ronald Reagan, voters in Washington, D.C. approved a basic right to shelter. The ballot initiative made the Black-majority and newly self-governing city one of the first jurisdictions in the country to legislate a right to shelter and, to this day, the only one to do so by popular vote (Greenberger, Brown, & Bowden, 1993; Roisman, 1991). In March 1985, when the law went to effect, a decent place to sleep each night was supposed to be available to anyone in the city who needed it. Instead, shelter space was limited and exposed residents to beatings, rats, lice, and scabies. After six years of laborious and multi-sited policy-making efforts by a range of actors – community activists, judges, lawyers, legislators, agency employees, and the mayoral administration – rights to these "virtual hell-holes" were undone (Williams, Jackson, Trinity, & Schorr, 1993, p. 57).

This paper traces the slow un-doing of the right to shelter law and explores what this un-doing means for understandings of urban governance. Here I turn to empirical data about the institutional mechanisms, political ideologies, social mobilizations, and geographical context through which the right to shelter was adopted, contested, implemented, and eventually overturned. Through a case study of the right to shelter, I contribute to the conversation about policy-failing in the burgeoning subfield of policy mobility studies (Baker & Temenos, 2015; Bok & Coe, 2017; He, Li, Zhang, & Wang, 2018; McCann & Ward, 2011; McFarlane, 2011; Peck & Theodore, 2015; Rutland & Aylett, 2008; Temenos, Baker, & Cook, 2018; Temenos & McCann, 2012). This paper explores the particular forces that made the right to shelter law move, and in many cases not move, within the city's institutional structures, discursive spaces, and built environment. It argues that research about how policies are un-made is a subject to be embraced rather than avoided. For geographical investigations of policy-making, there is much to be gained by close attention to the nitty-gritty work, networks, and sites of policy repeals.

Policy-failing in geography

Recent years have seen a flurry of work in geography around the topic of policy failure (Baker & McCann, 2018; Chang, 2017; Lovell, 2017a, 2017b; Malone, 2018; Wells, 2014, 2015). Largely emanating from the policy mobility subfield and its focus on "where, when, and by whom" policies are produced (McCann and Ward, 2010, 176), scholars have begun to explore the importance of failure to governance studies. For some, policy failure refers to a project that does not work well or a policy whose implementation did not go as planned (e.g. Peck, 2011; Peck & Theodore, 2010; Prince, 2010; Webber, 2015). Chang (2017) in this vein examines how a Chinese eco-city program that did not work well was still mobilized to influence policy elsewhere. For others, policy failure refers to the immobility of a policy. Stein, Michel, Glasze, and Pütz (2017) explore how the model of Business Improvement Districts has not been mobilized in Germany. Malone (2018) takes on the question of scale with regard to policy mobility in an investigation of residential security taxing districts that spread across but not beyond the city of New Orleans. For him, policy immobility at the inter-urban scale is not a fault line for failure. The policy after all was repeatedly transferred at the neighbor-hood level. Lovell's (2017a, 2017b) approach to policy failure encompasses both of these perspectives. She explores the circulation of failed policies, or those that have had unexpected results; *and* she examines the immobility of successful policies, or those that had expected results. For her study of an Australian smart metering program, failure describes both the policy outcomes *and* the mobilization processes. She suggests that how policies that have not worked well likely travel differently than those policies deemed best practices. The focus of her work is "what moves and what does not," be it the immobilization of successful policies or the mobilization of failed ones (2017a, p. 15). In my work the focus has been the moments – be them temporary, repeated, or permanent – in which a governance plan or project is defeated, stopped, or stalled. I have traced the interruptions to a plan to sell public property within one city and questioned the relationship between policy-making and policy-failing. To stress the processual and temporal nature of the on-going and incomplete set of practices, processes, and contexts through which a policy is made to fail, I employed the term *policyfailing* rather than policy failure (Wells, 2014). Instead of the mobilization of failed policies or the immobilization of successful ones, my research has taken up the narrower subject of the nitty-gritty processes and geographic contexts in which policy-making efforts are undone and made dormant.

The conceptual ambiguity about what exactly it means for a policy to fail lays a challenge for the expanding policy mobilities literature. By pointing out this slippage of terms, I am not calling for a categorization of failure types or a fortification of definitions about what really is a policy failure and what is not. I make this point about the various ways in which the term failure gets used as a way to demonstrate that the policy failure literature is in its infancy. For that reason, the next order of business should be an expansion of the scholarship's empirical scope. In particular, this paper calls for an expansion of the policy mobilities subfield to include the nitty-gritty practices and conditional forces through which policies are repealed. What has so far been overlooked is how some policies, be they mobile or immobile, are undone altogether. There is tremendous agreement in geography that policies are, as Stein et al. (2017, p. 37) note, "constantly reassembled" and "move[d] in bits and pieces." This notion of policy re-makings – detours, deviations, mutations, and complex amalgamations of places, scales, and actors – has become a mantra of sort for policy mobilities (see also McCann & Ward, 2015; Temenos & McCann, 2012; Ward, 2017). Baker, Cook, McCann, Temenos, and Ward (2016, p. 460), for instance, point out that "barriers and constraints are important features in the geographies and mobilities of policy." What has become important, as Chang (2017) suggests, is how policies – those "manufactured package[s] of different strands of knowledge of governance" (Bok & Coe, 2017, p. 53) – are held together and broken apart. On the methodological implications of this approach to policy, Robinson (2015, p. 7) writes: "Arriving at policies involves far more than assembling discrete materialized entities, ideas or objects which we can trace as they move from there to here. Complex, topological spatial imaginations are needed to interpret the mixing and folding of here" and multiple elsewhere. Policy mutations, multi-scalar forces, and tensions have been recognized as lenses into organizations and reproductions of power (Longhurst & McCann, 2016). But researchers have yet to fully consider how policies are un-assembled and undone altogether.

The empirical move that this paper makes helps to tease out a response to a pressing analytic question: What is the utility of the policy success and failure dichotomy (e.g. Davidson, 2017; McCann, 2017; McCann & Ward, 2015)? Critiques leveled against failure contend that it may be a futile framework, too ambiguous in its meaning. Ward (2018) suggests that it is "a moot point" to debate whether a particular policy rendition is a failure. In his study of a tax increment financing plan that was stalled for twenty years in Edinburgh, UK, he approaches the notion of failure primarily as a framework through which to assess policy consequences rather than as a political-geographic process to be investigated. Moore-Cherry and Bonnin (2018) similarly express skepticism that about the utility of failure as an analytic through which to understand an urban development project in Dublin, Ireland. They explore how time plays a critical role in whether a policy is deemed a success or failure. The authors, like Ward, put failure in quotes to remind readers of the critical lens through which the subject must be viewed. For Kassens-Noor and Lauermann (2018) study of an Olympic Games bid, the notion of a programmatic failure is something to be approached with caution. They write, "failure and success are moving targets that can be reframed over time" (Kassens-Noor & Lauermann, 2018, p. 3372).

Other scholars embrace the possibility of failure as, what Baker and McCann (2018, p. 5) write, is "an apt term to describe the reality of certain occurrences." In what can be read as a defense of sorts of policy failure studies, Baker and McCann (2018, p. 15) approach failure not as a "decontextualized, discrete, and functionally inert" policy-making event but as a geographic process seeped in a "social, political, spatial, and temporal context" and one that merits investigation (see also McCann, 2017). Set against the speed of Peck and Theodore (2015) work on fast policy, Baker and McCann (2018, p. 15) point out that the rejection of a supervised drug consumption site program in Melbourne, Australia may constitute a form of "slow policy-making." Longhurst and McCann's (2016) study of the politicized barriers that constrain the circulation of harm reduction drug policies builds on this idea of slow policy-making. Politicized barriers, which they call policy frontiers, lead to slow forms of governance. These recent theoretical contributions to the policy mobilities literature leave space for the possibility that some policies may be slowed to the point of eradication. Policy frontiers may be a step on the long and winding policy-failing road, which likely includes many detours of slow policy-making. By turning to the subject of policies that are no longer practiced and discussed – those that have failed the most and exist the least – the distinction between success and failure grows firmer and becomes less interpretive. The contribution that I make here is straightforward. I call for attention to repealed policies and the processes by which policies are un-made. In doing so, I offer a research trajectory for urban geography that broadens policy-failing as an empirical subject of inquiry. Policy-failing derives its strength, as this paper shows, not as an analytic framework but as an empirical object for investigation.

In the remainder of the article I analyze the rise and fall of the right to shelter law in Washington, D.C., drawing on archival research conducted in 2011 and 2012. For those years, I lived in the city. The richest trove of primary data for this project came from Gelman Library Special Collections at George Washington University. There I found letters and notes from two activists who organized the right to shelter ballot initiative (Mitch Snyder and Carroll Fennelly), as well as materials from the ballot initiative's opponents, namely the local Coalition for the Homeless and the city government. I also looked at the archives of the Luther Place Memorial Church, a church which provided homeless services in the 1980s and had a long-running and sometimes contentious relationship to the activists. I gathered media stories from national and local sources, including *The Washington Post, The Washington Afro-American, The Washington Times, The Washington Blade*, and the now defunct *DC Gazette*. This research draws on 25 transcribed, hour-long interviews that I conducted in-person with realtors and developers (6), planners and policymakers (6), lawyers (2), and housing advocates and experts (11) from the 1980s. I contacted the majority of these participants through snowball techniques or personal connections, as I had earlier lived in the city, conducted research there, and participated in local campaigns. In addition, I paid special attention to the D.C. government's court filings, publications, and public reports, and used these documents to explore how and why city officials opposed the right to shelter law. This data set is limited by the omission of the records of the D.C. Council hearings at which the evisceration of the right to shelter was debated and eventually approved. The microfiche tapes which held the transcripts of those hearings had been damaged by water in a storage center according to D.C. Legislative Services.

Historical context: shelter rights and the courts

Homelessness was nothing new in the USA when it surged in the 1980s (see Kusmer, 2003). Its face, however, was changing (see Blau, 1993; Hopper, 2003). In the preceding era of skid row and single occupancy residences, homeless populations had largely been made up of socially isolated, white, elderly men who were alcoholic, minimally employed, or unemployable altogether (Mitchell, 2011). In the 1980s, however, a host of factors – a shrinking housing supply, escalating interest rates, downward trends in real wages, social welfare cutbacks, and deinstitutionalization – diversified and drama-tically increased the homeless population to 3 million, many of whom were now women, children, and people of color (Rosenthal & Foscarinis, 2006). Federal policy-makers responded to this new surge of unsheltered residents by denying that it existed, treating it as a choice of individuals that did not merit government assistance, or suggesting that it was a temporary aberration rather than a predictable result of the country's economic and social policies. In 1981 the federal budget appropriated less than four percent to services for poverty reduction and alleviation but nearly sixty percent of budget cuts that year were made against programs that helped the poor (Kodras, 1997). Federal funds for community development, housing programs, and social services were cut in half in the first part of the decade (O'Connor, 2008). The provision of shelter services was mostly left to charities and, as a result, highly uneven across areas of the country (Langdon & Kass, 1985).

As a result of these fiscal moves and legislative inaction, the judicial branch became a place where activists often fought to secure resources for unsheltered residents. The threat of litigation, for instance, made New York City the first jurisdiction in the country to recognize state responsibility for shelter in 1979 (Hartman, 2006). Several other jurisdictions followed suit: the state of West Virginia; Los Angeles county in California; the counties of Orange, Westchester, and Dutchess in New York; and Atlantic City, New Jersey (Langdon & Kass, 1985). In this period the city of Philadelphia and the state of Massachusetts stood out as they added a right to shelter through the rare route of legislation (Roisman, 1991). "Without question, litigation" in the 1980s "proved to be an effective means for securing emergency relief" (Kim & Cox, 1986, p. 313). But it was a limited means. A homeless advocate from the period recounts: "As we saw the numbers increase and the faces change, we realized we needed to do advocacy at the local, state, and especially at the federal level," which required multi-city coordination.[1] In San Francisco in 1982, he and some of the activists who had brought the New York City case founded the National Coalition for the Homeless, an advocacy umbrella organization with chapters in cities like Washington, D.C. and a key network for the circulation of strategies and ideas about shelter rights.

In Washington, D.C. the 1978 election of Mayor Marion Barry, who would go on to be mayor for 12 years and who had first come to the city as chair of a Student Non-Violent Coordinating Committee (SNCC) chapter, marked a hopeful moment for local residents concerned about homelessness. After a hundred years of control by the federal government, city residents had a few years earlier earned the right to elect their own mayor and city council. Of the first Council, all but two of the thirteen members were black and the majority had ties to SNCC (Walters & Travis, 2010). "Nowhere else in the country," Asch and Musgrove (2017, p. 394) write, "had a black protest organization so thoroughly come to dominate a city government." City officials put racial justice and economic security at the top of their agenda (Paige & Reuss, 1983), and passed a series of anti-displacement and pro-tenant policies including rent controls, limited equity co-operative housing provisions (Huron, 2018), a speculation tax (Wells, 2015), and eviction restrictions (Greenberger et al., 1993). But, the city's fiscal conditions mounted a challenge to these policy priorities. The results of the Barry Administration's 1980 audit – the first in the city's history – were startling. The 630,000-person city had a $115 million deficit, a long-term debt of $300 million, and an unfunded pension liability of $750 million (Asch & Musgrove, 2017). The federal government's handover of limited self-governance to city residents had come with significant strings attached, strings that would eventually help to ensnare the racially progressive city in municipal bankruptcy in the mid-1990s.

In response to these fiscal conditions, Mayor Barry unveiled an austerity plan that fell most heavily on poor residents, nearly 15,000 of which were estimated to experience homelessness (Martinez, 1992). In April 1980 he announced that the budget shortfall was forcing him to close and sell three of the city's few homeless shelters within the next seventy-two hours. A local government agency director described Barry's efforts: "This time, the nobility of balancing the budget won over the nobility of community service or nonprofit use of the buildings." Almost immediately, a group of anarchist Catholic activists from the Community for Creative Non-Violence (CCNV) partnered with pro-bono lawyers to file suit and demand that the city government not close the shelters before replacement spaces were made available. CCNV had been the force behind the opening of these shelters. In 1978, after three men died of exposure to the cold, CCNV used street vigils, political theater, and acts of "disruptive mischief" (Hopper, 2003, p. 178) to convince the city government to re-appropriate vacant school buildings as shelters for the city's growing bloc of evicted tenants – more than 10,000 in 1975 alone (Bogard, 2003; Elwell, 2008; *The Washington Post*, 1976).

In court CCNV argued that shelter benefits constituted a kind of property right that could not be taken away without due process and compensation (Williams et al., 1993). The court disagreed and said that the city's decision to liquidate property assets was not a subject for judicial review:

> Considering the purely political nature of the shelter closing decision, and that the fact that there are no requirements, statutory or otherwise, which the District's ultimate decision can be measured against, this court agrees with the defendants that the only review that can be possibly conducted is a review to ensure that the District followed the proper procedures. (quoted in Langdon & Kass, 1985, p. 324)

A federal judge also said "the government assumes no obligation to house and feed indigent people" (quoted in *The Washington Post*, 1980). After an appeal, a federal judge agreed that there is not "a constitutional or other legal right to city-provided shelter" (quoted in Langdon & Kass, 1985, p. 324). The courts did, however, say that the local government had created a temporary entitlement to shelter for residents already in shelters and violated the rights of those residents when it did not give residents advance notice of shelter closures. There was a legal basis, in other words, not to lose one's existing shelter access without reasonable notice or due process though the courts maintained that there was not a legal basis for city residents to always be offered shelter. For CCNV the important point was this: The courts did not say that the guarantee of shelter in the city was unconstitutional or one that could not be upheld at some later point. The courts simply said that such an entitlement did not currently exist.

This ruling set in motion CCNV's campaign to establish a right to shelter through a popular vote. This ruling also foreshadowed the extent to which the making and failing of the right to shelter law would be slow, laborious, and stretched across various sites of power including the judiciary, the executive and legislative branches, the media, and the streets. Washington, D.C.'s right to shelter law, from its very inception in the courts in 1980, was a seed whose roots never grew deep enough. The following sections document all the winds, rainstorms, and droughts that made the city's soil so inhospitable to that seed of a right to shelter. This paper documents how one policy was slowly and laboriously unearthed from a multitude of sites in a single city over the course of ten years.

The initial vote

In fall 1983 CCNV moved forward with a plan to put the question of shelter rights on the November 1984 ballot (Rader, 1986). The activists secured approval from the D.C. Board of Elections and Ethics to circulate a petition and, as a CCNV member described, spent three months collecting signatures:

> We were all over the city... Half the day we were spending doing filling out the petitions. The rest of the time people were down at the Board of Elections verifying that these signatures were real and not dupes. Oh it was a nightmare. I mean, we were out all day long. You know, people from the community house would go. Folks from the shelter would go out. You sometimes go to Georgetown, and go sit in front of the "Social" Safeway [a grocery store] and be there eight, nine hours.[2]

CCNV's team of 85 or so people collected more than 32,000 signatures (*The Washington Post*, 1986). It was almost a third more than was required. Part of this success was due to the fact that it was a presidential election year and CCNV was able to fill out its petitions during the May 1984 primaries. The petition also garnered attention and support through a secondary issue about whether homeless residents had the right to vote. After five voter application cards were rejected, CCNV allies sued, won, and made Washington, D.C. the first jurisdiction in the country to guarantee homeless residents the right to vote (*The New York Times*, 1984).

Six weeks before the November 1984 vote on "The Right to Overnight Shelter Initiative 17," Mayor Barry sued the D.C. Board of Elections and Ethics to try to prevent the initiative from appearing on the ballot. Even though a few months earlier Mayor Barry had signed CCNV's petition and helped CCNV to secure a federal property for use as a shelter, his administration now took the position that a mandate for overnight shelter was an improper subject for a ballot issue. Barry's administration argued that the city's charter did not allow voters to pass laws that could have funding implications. When the court disagreed and the issue was cleared to appear on the ballot, Barry and his allies pursued a series of different strategies, to which we now turn, to the fight against the right to shelter initiative.

Barry's supporters warned the public that any expansion of a public shelter system could absolve the federal government of its responsibilities for housing and homeless services, unfairly burdening the local coffers. The local chapter of the National Coalition for the Homeless stressed the scalar problems that such a law could pose: "Why should we support the federal government in its un-Christian and unethical move to relinquish its responsibility to the homeless?"[3] The local chapter was often quoted in media coverage and public debates and served as an important oppositional figure, a source of outrage for the national organization. This opposition can be attributed to two factors: close personal relationships with Mayor Barry, and dislike for CCNV's leaders who had not sought the input of the shelter provider community on the ballot issue. An expert on homelessness explained:

> There ended up being some rifts in the community... There was a movement of concerned members of the Coalition, of which I was one, to try to keep the Coalition from going in that direction... There was CCNV on the one hand, and then there were the people who 'played by the rules' on the other hand.[4]

The local chapter of the Coalition for the Homeless argued that the right to shelter could warehouse people by expanding the shelter system and reduce resources for affordable housing and transitional housing efforts. The D.C. Commission on Homelessness, an advisory board to the mayor, agreed. In an op-ed in *The Washington Post* (1984), the local Coalition said, "A No Vote is a vote that no homeless person will be homeless forever." A letter sent to the local Coalition by a doctor disagreed with this point: "I have not heard the board saying that soup kitchens should be closed because homeless people deserve better."[5]

Mayor Barry and his allies also made arguments that the right to shelter would bankrupt the city and force high taxes onto residents. *The Washington Afro-American* (1984) predicted that the city would become "a haven for the homeless of the world." *The Washington Post* (1984) printed an editorial that described the initiative as "an invitation to people of any means to demand...shelter for what could be the rest of their lives." Wealthy people would want the free overnight accommodations in the shelters, city officials claimed, and flood the city. To spread these claims, the Barry administration illegally spent $7,000 worth of public funds to mail 50,000 yellow pamphlets to city workers and put up 600 red and white posters around the city. And, it illegally assigned on-duty city employees to poll stations to urge a "no" vote (Williams et al., 1993).

What becomes apparent in the archives and interviews with activists and policy-makers at the time is the extent to which the campaign was imbued with multi-scalar tensions from its inception. Although Mayor Barry fought against the ballot initiative in 1984, and subsequently resisted its implementation for the next six years, he often stood alongside CCNV during the same period on other matters to fight for *federal* provision of housing and shelter. For the Barry administrations, the question was often about who would foot the costs, not whether a particular entitlement program should be established. In 1985 he declared February to be "Friends of the Homeless Month" and announced a fund drive to raise $150,000 for emergency homelessness assistance. He said: "Now is our chance to vote with our wallets" (quoted in *The Washington Post*, 1985). In 1987 he stood with CCNV to ask the federal government to fund renovations to a shelter located in a federal building (Bogard, 2003) and, in 1989, held fundraisers for the national Housing Now March, which brought together CCNV supporters and 250,000 other homeless advocates. One of the March organizers described the alliance:

> I remember being in a meeting…a lunch fundraiser. [Mayor Marion Barry] brought in his developer friends, his lawyer friends, and they all pledged something like $2,000 to $5,000, which was huge money at that time…. We weren't attacking the city. We were attacking the federal government for not guaranteeing a right to shelter.[6]

This interview excerpt showcases the jurisdictional fault line between the city government and federal government on which the right to shelter law was built.

By the time of the vote in November 1984, CCNV had amassed broad support. It had distributed 150,000 sample ballots, but its power came from its fifteen years of local and national activism (Rosenthal & Foscarinis, 2006). In Washington, D.C. CCNV ran a soup kitchen, a drop-in center, and a free food store. That year its members released cockroaches into the White House, set-up a tent city called Reaganville in front of the White House, and hosted U.S. Congressional hearings on homelessness in its shelter. CCNV had also landed a story on the cover of *Newsweek*, organized "Voices from the Streets," an event at which Martin Sheen narrated homeless people's stories and Pete Seeger played music on the steps of the U.S. Capitol, and argued before the U.S. Supreme Court that sleeping in tents constituted a protected form of political speech. "What other group," a CCNV member asked, "had every branch of the U.S. government responding to their claims?" (quoted in Rader, 1986, p. 216). The apex of CCNV's activism in 1984 was a vow to fast until the federal government agreed to renovate a federal building in Washington, D.C. in which CCNV sheltered 850 people each night. After CCNV's most prominent member, Mitch Snyder, fasted for 51 days, federal officials agreed to the demand. That night a *60 Minutes* television segment covered the story. The timing of the fast no doubt helped to ensure the passage of the right to shelter initiative three days later, but the ballot issue was by no means the primary focus of CCNV's work. It was, as a CCNV member said in an interview, "one of the strategies we used to gauge the level of compassion and sympathy to the homeless in the city;" it was a stepping-stone.[7] On 6 November 1984 seventy-two percent of voters in Washington, D.C. approved the right to shelter ballot initiative, which read:

> The District of Columbia, in recognition that: 1) All persons have a right at all times to overnight shelter adequate to maintain, support and protect human health; 2) The costs of providing adequate and accessible shelter to all in need are outweighed by the costs of increased police protection, medical care, and suffering attending the failure to provide adequate shelter; and 3) It is in the best interest of the District to provide overnight shelter for the homeless; Hereby establishes in law the right to adequate overnight shelter, and provides for identification of those in need of shelter and provision of such shelter (quoted in Martinez, 1992, p. 9).

Every ward in the city voted yes. The first vote of any kind in the USA on the issue of homelessness was a success. Four months later, in March 1985, the voter-created right to shelter became law. It would only be law, however, intermittently over the next five years before its nullification. The Barry administrations repeatedly challenged the initiative in court, sometimes successfully. From July 1985 until May 1986, for instance, a court agreed with the local government's claims that the subject was an improper one for a vote and invalidated the right to shelter. This gap year ended when CCNV won an appeal. What did not end was the city government's ongoing resistance to the law, refusal to implement it, and efforts in the courts to revoke it (Reid, 1986). The right to shelter, which had begun in the courts and moved to the ballot, would stretch across a series of governance spaces including the mayoral office before its undoing by the D.C. Council.

Government resistance and a referendum

A month before the right to shelter went into effect, city officials closed the only shelter in the wealthy Georgetown neighborhood. A month after the law went into effect, city officials boarded up another 400-bed shelter. By the end of 1985 there were only 100 more beds in the city than there had been before the passage of the right to shelter (*The Washington City Paper*, 1989). The shelters were decrepit and mired in the corruption that characterized the newly self-governing city. City officials, for instance, spent millions of dollars to shelter residents in vermin-infested hotels that were owned by businessmen with ties to the mayor (Asch & Musgrove, 2017; Fauntroy, 2003).

In 1988 the city government's refusal to implement the right to shelter law kindled two lawsuits. The first sought to stop public auctions of the vacant Lennox and Morgan schools to high-end condominium developers. In efforts to balance its budget, the local government in the 1980s had undertaken a selling spree of public buildings, some of which were in-use as shelters (Jaffe & Sherwood, 1994; McGovern, 1998). As it had in 1980, the court ruled against CCNV on the question of whether public property should be prioritized for unsheltered residents. CCNV made headway with the second lawsuit, *Atchison v. Barry*, which found the city government was not upholding the right to shelter. Similar to legal actions in New York City, the DC courts ordered months of interviews and investigations by a ten-member team of lawyers, law students, priests, and homeless activists. The investigations, and subsequent ones over the next few years, found that the city had failed to implement the right to shelter. City officials refused to open enough shelters to meet demand, ignored qualitative standards, disregarded requirements to publicize shelter locations, failed to provide security for shelter residents, misused federal grants, and created multi-million dollar scandals with contract cronyism (Williams et al., 1993). A pro-bono lawyer involved in the cases stressed that this resistance in the 1980s was not unique to the right to shelter law. He said, "the government resisted as much as possible" compliance with laws about foster care, nursing homes, and drug rehabilitation facilities.[8]

Eventually, the court banned the city from closing any of its shelters and ordered that the city add fifty shelter beds to its supply when the existing supply had a ninety-five percent occupancy rate for six days in a row. In response, city officials took a shortcut and set up mobile trailers in government-owned parking lots in the Northeast neighborhoods of Trinidad and Ivy City. Katherine Boo, then a journalist for the local weekly, *The Washington City Paper* (1989), explained:

> With 18 beds each, the trailers are small, relatively cheap, and easier than a gymnasium to keep quiet and clean... When necessary, city officials can simply roll in another 18 beds... Shelters like these are the wave of the future in the District. From 7 p.m. to 7 a.m., on beds in trailers behind chain-link fences, men and women will exercise their 'right to shelter.'

City officials also put cots in the gyms of in-use schools, a psychiatric hospital, a rooming house with no heat and broken windows, the pro-football stadium, and the D.C. Council building. A longtime city official who was working in the Council building at the time described the conditions:

> We would leave the building at 6 o'clock. They were setting up cots on the ground floor... There would be enough homeless people down here that they would line the walls on the ground floor... They worked out an arrangement with the security officers so that...you could get through the doors, in and out of the buildings, single file. And you could go through certain passages, single file. And there were just people everywhere. And they just sat, and sat, and sat.[9]

There were more than 36 locations across the city, including the trailer parking lots, where roughly 18,000 individuals accessed overnight emergency shelter in 1989 (Henig, 1994). These numbers could seem like a victory, if not a perverse one, for homeless residents and advocates. In the year before the passage of the right to shelter the public shelter system only served about 2,000 individuals (Henig, 1994). But the landscape of provisions that the city built through these trailers was a temporary one, and intentionally so. Interviews and records in the archives confirm that city officials had little intention of complying with the existing law. The trailers were a stopgap measure until the law could be undone. The trailers also posed a barrier to the law's materiality. The physical landscape of shelter provisions was easy to disassemble.

When the city refused to improve the conditions of shelters, the court began to issue daily fines of up to $30,000. City officials in the Barry administration opted to pay these penalties rather than spend a portion of the fees to bring the shelter system into compliance. Although the Chairman of the D.C. Council, David Clarke, disagreed with this tactic, other members of the local legislature voiced support for an end to the right to shelter law. Councilmember Nadine Winter introduced an amendment in 1989 to limit shelter use to ten days per six-month period and require proof of residency. She said, "We have done what we can to make them human beings."[10] Another Councilmember agreed: "I don't believe everybody has a right to shelter." Outside of D.C. Council, in a neighborhood where several temporary trailers had been placed, residents drafted a voter referendum to repeal the right to shelter. A crack epidemic had taken center stage and further strained public budgets (whose debt would reach $722 million by 1995 and propel a federal takeover of the city).[11] A staff member of the D.C. Council described in an interview the extent to which the public deficit pressed on its political climate: "Most everything back then was being driven by finances."[12] Councilmember Frank Smith echoed this notion in an interview: "For us, to do something like this, it had to be self-funded and self-financed, otherwise you had to rob people to pay Paul because the government didn't have the money."[13]

In the end CCNV's success in the courts became a Faustian bargain. The court fines, which had grown to $4.4 million, were mobilized in public discourse as the central rationale for the legislative cuts to the right to shelter. The law was too costly, legislators said. In June 1990, after a heated three-hour debate, the D.C. Council approved an amendment to limit a whole host of factors about the right to shelter law: the amount of funds that could be spent on shelters, the number of nights that eligible individuals could use shelters each year, and the months when shelter would be offered (hypothermia season). The D.C. Emergency Overnight Shelter Amendment Act also eliminated the requirement for dignity in shelters and set-up a number of eligibility requirements. Residents with suspected addiction problems would be required to participate in treatment programs while those who were deemed capable of working would be required to pay a fee for shelter access or perform community service. Most notably, shelter would be granted only to residents *without* a credit card – presumably those who were unable to take on consumer debt – and those who had regularly paid income taxes (Williams et al., 1993).

In an attempt to overturn the D.C. Council's amendment and reinstate the original right to shelter, CCNV organized a voter referendum for the November 1990 ballot. An estimated 500 CCNV members and volunteers raised $44,000 and collected a record 43,000 names, achievements that suggested the measure would pass easily.[14] A poll of residents confirmed broad support for the referendum with 71 percent of participants reporting that they would vote for the original right to shelter law (Williams et al., 1993). Still, an expert involved in the campaign said: "It was kind of a horrible time. Mitch Snyder [CCNV's most well-known member] had just died and the community was kind of in upheaval. So some of the advocacy to get the referendum passed was compromised."

As it was at the time of the first vote on a right to shelter, the largest force mobilizing against CCNV's referendum was the city government. The Barry administration again illegally used public funds to print and distribute materials that encouraged residents to reject CCNV's ballot referendum. This time, however, there were two key differences. First, the city government framed homelessness as a problem of individuals (rather than as a societal problem) and treated shelter as an overly generous entitlement (rather than a necessary program for the federal government to fund). A government-produced flyer said, homeless people must "accept help for the problems that caused their homelessness."[15] Second, local legislators were invested in the battle and voiced strong opposition to the referendum. It was, after all, their amendment that was in question. Councilmember H.R. Crawford wrote an editorial in *The Washington Post* (1990):

> Why should one person's right to shelter come at the expense of our other residents? What about programs for the mentally retarded, adequate books and laboratory equipment for our schools, increased pay for our teachers, an adequate level of homemaker and chore services for senior citizens and others to prevent institutionalization, proper staffing and equipment for our public health clinic...

Before the referendum vote Mayor Barry was arrested for drug use. He had been seeking re-election in November 1990 for a fourth term. When he dropped out of the race, his challenger, and eventual successor, Sharon Dixon took the same position as Barry on the issue of the right to shelter that it was costing the city too much money. According to a homelessness expert, Dixon also said the law "was attracting people from other parts of the country."[16] This was a common refrain that the right to shelter law invited other jurisdictions to send their homeless residents to Washington, D.C. There appears to have been some truth to this claim but, as an activist noted in an interview, people mobilized the story of Boston buying a family a one-way bus ticket to Washington, D.C. "as an excuse to get rid" of the right to shelter law.[17] To this day, the story still circulates. The director of a business improvement district in an interview question about the right to shelter made this comment: "How many people moved here from Kentucky? When you're out of line with everybody else in the region, you can attract things, you know."[18]

On 6 November 1990, almost six years to the day that the right to shelter first appeared on the ballot, voters in Washington, D.C. narrowly defeated CCNV's "Referendum #005" to reinstate the original right to shelter. Of the 125,000 votes cast on the issue, the issue lost by 3,179 votes, just two percent. Whereas the initiative to establish the right to shelter passed in all areas of the city, the geography of the referendum results differed dramatically. There was a strong pattern of race and wealth in the distribution of votes. There was clear support for the measure in areas of the city with significant populations of black and poor residents (Henig, 1994). By contrast, voters struck down the referendum by almost a 2 to 1 margin in the wealthy, home-owning, and predominantly white neighborhoods of the upper Northwest quadrant. Of the city's eight wards, the referendum failed in the three wealthiest and whitest ones, which included the gentrifying neighborhoods of Capitol Hill and Dupont Circle. Despite this pattern, several supporters of the referendum relayed to me in interviews that the results of the referendum were an accident having to do with messy grammar. The referendum was worded in such a way that some people who supported shelter rights were thought to have accidentally voted against it.[19] The referendum text was as follows (quoted in Williams et al., 1993, p. 71):

> Referendum #005 rejects the 1990 Act that changes the current Right to Overnight Shelter Law ("Initiative 17"). The 1990 Act changes the current law ("Initiative 17") by: 1) Removing the entitlement "right" of homeless persons to overnight shelter; 2) Establishing a program (but no entitlement) for shelter and support services for the homeless; 3) Defining eligibility for receiving shelter, grounds for denying shelter, limits on the length of stay, participation in costs by shelter occupants, and other requirements.
>
> Vote "FOR" Referendum #005 to reject the 1990 Act and retain the current law.
>
> Vote "AGAINST" Referendum #005 to permit the 1990 Act to become law.

To reject the D.C. Council's amendment, voters had to *approve* the referendum, which was counter-intuitive. A lawyer who was involved with the campaign said "people didn't really understand which way to vote." A study conducted a month after the election supported this possibility when it found that three-fourths of 400 surveyed residents believed that the city government was not doing enough to help homeless residents and that homeless people should be "guaranteed a safe place to sleep" (Williams et al., 1993, p. 73).

When the new limits on the right to shelter went into effect in March 1991, the city government had already begun to approach the right to shelter law as a relic. In the months prior, city agencies sold three shelters and reduced shelter capacity by 800 beds (Greenberger et al., 1993). Enabled by the legislature and the voters, and no longer entangled in the compliance constraints of the courts, the executive branch of the local government moved closer to the finish line of its multi-year and multi-sited marathon to undo the right to shelter law. City officials have suggested in interviews that the problem with the right to shelter law was not implementation but the actual goal to offer shelter to anyone who needs it every night of the year. A current councilmember, who was active in local policy in the 1980s, said that the right to shelter law, if proposed today, would not pass:

> I think the city has matured. I mean, the right to shelter was..very broad brush liberal. 'We've got home rule now and we can act like San Francisco, or we can act like Takoma Park, you know, which...is a nuclear-free zone.... We're going to strike these blows for a better society.' And I don't want to knock that concept. But... [the 1980s policymakers] were a little bit too simplistic.[20]

The failing of the right to shelter has been blamed on the merits and logic of the policy – not on the messy array of actual contexts, paradigms, and practices that shaped its implementation and short life, namely the city's debts, its sales of public property, its use of trailers, the court fees, the fuzzy referendum language, federal government relations, and, perhaps most importantly, local government dysfunction.

Conclusion

This paper contributes to critical policy studies literature by developing the concept of policy-failing and demonstrating some of the actual contexts and micro-practices involved in the case of a policy repeal. The case of the right to shelter showcases the utility of thinking about the political-geographic processes by which policies are undone, overturned, and, in certain situations, repealed. Policy repeals are the deleted paragraphs and architectural ruins of oft-forgotten policy-making struggles. By excavating ten years of a policy-failing, this paper offers insight into the nitty-gritty work, networks, mobilities, and timetables through which cities get governed.

The right to shelter law was a de facto skeleton in the 1990s. In 2005 local policymakers officially removed from the legal code all parts of the Right to Overnight Shelter Act of 1984 and adopted the new Homeless Services Reform Act, which remains in place today and governs access to shelter during hypothermia season. Footprints from the original right to shelter law remain, especially with regard to the requirement for families to received apartment-style accommodations. Last year the city spent $25 million to shelter families in motels. The temporary landscape of shelter provisions today is no longer trailers but motels (WAMU, 2018). On days when the temperature is below freezing, temporary shelters are opened. When the temperature rises, shelters close and hundreds of people are left without access to overnight shelter. In 2007, for instance, the city closed six hypothermia shelters after the weather warmed and put 300 people on the streets. Sometimes the same buildings re-open the following winter; other times they are sold. Since the city's closure of three shelters in 1980, policymakers have repeatedly tried to close and sell public shelters to private developers of condominiums, gyms, and other high-end facilities. Between 1998 and 2014, for instance, the city tried to sell 68 percent of its 22 public shelters (Wells, 2014). Today the right to shelter offers a record of place-making struggles toward alternative urban futures.

This study supports Robinson's (2015) observation that policy-making efforts mix and fold together a host of conditions and forces, including institutional mechanisms, political ideologies, social mobilizations, and regulatory practices. The failing of the right to shelter was a protracted, laborious, and multi-sited political-geographic process. The threads of the fraying right to shelter were pulled from a number of sources and over a number of years. A close examination of the right to shelter's empirical data does not reveal a singular moment, site, or mechanism by which the right to shelter was made to fail. Instead the failing of the right to shelter emerged from an array of places of power, including the judiciary, the local executive and legislative branches, the media, and the streets There is no identifiable push that took down the whole chain of dominos. That being said, the dominos were made to fall and such a fall was not inevitable.

The case of the right to shelter demonstrates how the notion of failure is, as Longhurst and McCann (2016) suggest, a useful description of certain governance processes rather than an analytic distraction. Some policies, as this paper shows, do actually experience mutations and transformations to such a degree that they barely exist (apart from the life of an artifact). By turning to the subject of policies that are no longer practiced and discussed – those that have failed the most and exist the least – the distinction between a policy success and failure grows firmer and becomes less interpretive. Policy-failing derives its strength not as an analytic framework but as an empirical object for investigation.

The critical policy studies literature draws strength from its examination of mutations and transformations, but it has yet to fill out study of what role policy-failing – in particular policy repeals – play in particular contexts, whether policy-failing is intrinsic to all regulatory schemes, and what challenge policy-failing poses to existing governance regimes. If the concept of policy-making as full of mutations and mobilities is to be taken seriously, it will require attention to instances of how policies are un-assembled and undone. To better understand the production of uneven urban geographies, like those that persist in Washington, D.C., geographers must investigate the unevenness of policy-making. Such a research agenda requires attention to the slow, laborious, often contradictory, power laden, and spatial processes of how policies like the right to shelter are removed from a city's landscape.

Notes

1. Interview of homeless activist "F" with author, Washington, D.C., 22 March 2011.
2. Interview of CCNV member "G" with author, Washington, D.C., 26 March 2011.
3. Meeting Minutes dated 19 September 1984 of the Coalition for the Homeless. Carol Fennelly Personal Papers, Box 28, Folder 13, Gelman Library Special Collections, George Washington University in Washington, D.C. (Hereafter Fennelly PP).
4. Interview of legal homeless expert "A" with author, Washington, D.C., 24 March 2011.
5. Personal letter dated 26 September 1984 (Fennelly PP, Box 28, Folder 11) .
6. Interview of activist "C" with author, Washington, D.C., 18 March 2011.
7. Interview of CCNV member "G" with author, Washington, D.C., 26 March 2011.
8. Interview of legal homeless expert "B" with author, Washington, D.C, 26 October 2011.
9. Interview of city official "E" with author, Washington, D.C., 20 October 2011.
10. CCNV Flyer (Fennelly PP, Box 28, Folder 16).
11. Between 1985 and 1987, one in four males between the ages of 18 and 29 in D.C. were arrested on drug-related charges (Asch & Musgrove, 2017).
12. Interview of city official "E" with author, Washington, D.C., 20 October 2011.
13. Interview of city official "F" with author, Washington, D.C., 21 November 2011.
14. By contrast, an opposition group raised just $2,000, mostly from the Washington, D.C. Association of Realtors.

15. "Helping Homeless People Through Difficult Times" two-page flyer from the Government of the District of Columbia from 1990, Jointly prepared by the City Administrator, Deputy Mayor for Economic Development, the Departments of Human Services, Public Housing, Housing and Community Development, and the office of Communications (Fennelly PP, Box 28, Folder 13).
16. Interview of homeless activist "F" with author, Washington, D.C., 22 March 2011.
17. Interview of activist "D" with author, Washington, D.C., 7 April 2011.
18. Interview of business improvement district director with author, Washington, D.C., 26 October 2011.
19. Interview of legal homeless expert "A" with author, Washington, D.C., 24 March 2011.
20. Interview of city official "E" with author, Washington, D.C., 20 October 2011.

Acknowledgments

I would like to thank Cristinia Temenos, John Lauermann, Kate Coddington, Don Mitchell, Emily Billo, Kafui Attoh, and three anonymous reviewers for valuable feedback.

Disclosure statement

No potential conflict of interest was reported by the author.

References

Asch, Chris Myers, & Musgrove, George Derek (2017). *Chocolate City: A history of race and democracy in the nation's capital*. Chapel Hill, NC: UNC Press.
Baker, Tom, Cook, Ian R., McCann, Eugene, Temenos, Cristina, & Ward, Kevin (2016). Policies on the move: The transatlantic travels of tax increment financing. *Annals of the American Association of Geographers, 106*(2), 459–469.
Baker, Tom, & McCann, Eugene (2018). Beyond failure: The generative effects of unsuccessful proposals for Supervised Drug Consumption Sites (SCS) in Melbourne, Australia. In *Urban geography* (pp. 1–19). doi:10.1080/02723638.2018.1500254
Baker, Tom, & Temenos, Cristina (2015). Urban policy mobilities research: Introduction to a Debate. *International Journal Urban Regional, 39*, 824–827.
Blau, Joel (1993). The visible poor: Homelessness in the United States. *Administration in Social Work, 17*(4), 127.
Bogard, Cynthia (2003). *Seasons such as these: how homelessness took shape in America*. Hawthorne, NY: Aldyne de Gruyter.
Bok, Rachel, & Coe, Neil M (2017). Geographies of policy knowledge: The state and corporate dimensions of contemporary policy mobilities. *Cities, 63*, 51–57.
Chang, IChun Catherine (2017). Failure matters: Reassembling eco-urbanism in a globalizing China. *Environment and Planning A, 49*(8), 1719–1742.
Davidson, Mark (2017 April 6). *Discussant for "The urban politics of policy failure I: Political impacts of failure"*. Presentation at the Annual Meeting of the American Association of Geographers in Boston Massachusetts.
Elwell, Christine (2008)., From political protest to Bureaucratic service: The transformation of homeless advocacy in the nation's capital and the Eclipse of political discourse. Unpublished PhD thesis, American University
Fauntroy, Michael (2003). *Home rule or house rule?* Maryland: University Press of America.
Greenberger, M., Brown, E., & Bowden, A. eds. (1993). *Cold, harsh, and unending resistance: The district of Columbia government's hidden war against its poor and homeless*. Washington, D.C.: Washington Legal Clinic for the Homeless. Available online at www.legalclinic.org (last accessed 1 June 2015)

Hartman, Chester (2006). The case for a right to housing. In R Bratt, C Hartman, & M Stone (Eds.), *A right to housing: Foundation for a new social agenda* (pp. 177–192). Philadelphia, PA: Temple University Press.

He, Shenjing, Li, L., Zhang, Y., & Wang, J. (2018). A small entrepreneurial city in action: Policy mobility, urban entrepreneurialism, and politics of scale in Jiyuan, China. *International Journal Urban Regulation Research*, *42*, 684–702.

Henig, Jeffrey (1994). To know them is to…? Proximity to shelters and support for the homeless. *Social Science Quarterly*, *75*(4), 741–754.

Hopper, Kim (2003). *Reckoning with homelessness*. Ithaca, NY: Cornell University Press.

Huron, Amanda (2018). *Carving out the commons: Tenant organizing and housing cooperatives in Washington*. DC: University of Minnesota Press.

Jaffe, Harry, & Sherwood, Tom (1994). *Dream city: Race, power and the decline of Washington, D. C.* New York: Simon and Schuster.

Kassens-Noor, Eva & Lauermann, John (2018). Mechanisms of policy failure: Boston's 2024 Olympic bid. *Urban Studies*, *55*(15), 3369–3384.

Kim, Hopper, & Cox, L. Stuart (1986). Litigation in advocacy for the homeless: The case of New York City. In J Erickson & C Wilhelm (Eds.), *Housing the homeless* (pp. 301–314). New Brunswick, NJ: Center for Urban Policy Research.

Kodras, Janet (1997). The changing map of American poverty in an era of economic restructuring and political realignment. *Economic Geography*, *73*, 67–93.

Kusmer, Kenneth (2003). *Down and out, on the road: The homeless in American history*. New York: Oxford University Press.

Langdon, James, & Kass, Mark (1985). Homelessness in America: Looking for the right shelter. *Columbia Journal of Law and Social Problems*, *19*, 305–392.

Longhurst, Andrew, & McCann, Eugene (2016). Political struggles on a frontier of harm reduction drug policy: Geographies of constrained policy mobility. *Space and Polity*, *20*(1), 109–123.

Lovell, Heather (2017a). Policy failure mobilities. *Progress in Human Geography*, *43*(1), 46–63.

Lovell, Heather (2017b). Are policy failures mobile? An investigation of the advanced metering infrastructure program in the State of Victoria, Australia. *Environment and Planning A*, *49*(2), 314–331.

Malone, Aaron (2018). (Im) mobile and (un) successful? A policy mobilities approach to New Orleans's residential security taxing districts. In *Environment and planning C: Politics and space*, *37*(1), 102-118.

Martinez, S. Lynn (1992). An American vision: The right to shelter. *Buffalo Public Interest Law Journal*, *12*, 1–22.

McCann, Eugene (2017). Mobilities, politics, and the future: Critical geographies of green urbanism. *Environment and Planning A*, *49*(8), 1816–1823.

McCann, Eugene, & Ward, Kevin (2010). Relationality/territoriality: Toward a conceptualization of cities in the world. *Geoforum*, *41*(2), 175–184.

McCann, Eugene, & Ward, Kevin (2011). *Mobile urbanism*. Minnesota: University of Minnesota Press.

McCann, Eugene, & Ward, Kevin (2015). Thinking through dualisms in urban policy mobilities. *International Journal of Urban and Regional Research*, *39*(4), 828–830.

McFarlane, Colin (2011). *Learning the city: knowledge and translocal assemblage*. Malden, MA: Wiley-Blackwell.

McGovern, Stephen (1998). *The politics of downtown development*. Lexington: University Press of Kentucky.

Mitchell, Don (2011). Homelessness, American style. *Urban Geography*, *32*, 933–955.

Moore-Cherry, Niamh, & Bonnin, Christine (2018). Playing with time in Moore street, Dublin: Urban redevelopment, temporal politics and the governance of space-time. *Urban Geography*. doi:10.1080/02723638.2018.1429767

O'Connor, Alice (2008). Swimming against the tide. In J DeFilippis & S Sagegert (Eds.), *The community development reader* (pp. 29–47). New York: Routledge.

Paige, Jerome, & Reuss, Margaret (1983). *Safe, decent and affordable: Citizen struggles to improve housing in the district of Columbia.* 1890–1982. Washington, D.C.: University of the District of Columbia.

Peck, Jaime (2011). Geographies of policy: From transfer-diffusion to mobility-mutation. *Progress in Human Geography, 35,* 773–797.

Peck, Jaime, & Theodore, Nik (2010). Mobilizing policy: Models, methods, and mutations. *Geoforum, 41,* 195–208.

Peck, Jamie, & Theodore, Nik (2015). *Fast policy: Experimental statecraft at the thresholds of neoliberalism.* Minneapolis, MN: U of Minnesota Press.

Prince, Russell (2010). Policy transfer as policy assemblage: Making policy for the creative industries in New Zealand. *Environment and Planning A, 42,* 169–186.

Rader, Victoria (1986). *Signal through the flames.* Kansas City, MO: Sheed and Ward.

Reid, Inez Smith. (1986). Law, politics and the homeless. *West Virginia Law Review, 89*(1), 115–148.

Robinson, Jennifer (2015). 'Arriving at' urban policies: The topological spaces of urban policy mobility. *International Journal of Urban and Regional Research, 39*(4), 831–834.

Roisman, Florence (1991). Establishing a right to housing: A general guide. *Clearinghouse Review, 25*(3), 203–226.

Rosenthal, Rob, & Foscarinis, Maria (2006). Responses to homelessness: Past policies, future directions, and a right to housing. In R. Bratt, M. Stone, & C. Hartman (Eds.), *A right to housing: Foundation for a new social agenda* (pp. 316–339). Philadelphia: Temple University Press.

Rutland, Ted, & Aylett, Alex (2008). The work of policy: Actor networks, governmentality, and local action on climate change in Portland, Oregon. *Environment and Planning D, 26,* 627–646.

Stein, Christian, Michel, B., Glasze, G., & Pütz, R. (2017). Learning from failed policy mobilities: Contradictions, resistances and unintended outcomes in the transfer of "business improvement districts" to Germany. *European Urban and Regional Studies, 24*(1), 35–49.

Temenos, Cristina, Baker, Tom, & Cook, Ian R. (2018). Inside mobile urbanism: Cities and policy mobilities. In T. Schwanen (Ed.), *Handbook of urban geography.* Cheltenham: Edward Elgar.

Temenos, Cristina, & McCann, Eugene (2012). The local politics of policy mobility: Learning, persuasion, and the production of a municipal sustainability fix. *Environment and Planning A, 44*(6), 1389–1406.

The New York Times. (1984 August 29). *A right to live, a right to vote.* https://www.nytimes.com/1984/08/29/us/a-right-to-live-a-right-to-vote.html

The Washington Afro-American (1984 October 27). *Vote "no" on initiative 17.* Available in the Washingtonian Special Collections, Martin Luther King, Jr. Public Library, Washington, D.C.

The Washington City Paper (1989 September 29) Homeless economics. Available in the Washingtonian Special Collections, Martin Luther King, Jr. Public Library, Washington, D.C.

The Washington Post. (1976 December 8). Empty houses and the homeless poor.

The Washington Post (1980 May 24). City backs off its plan to lower health funds for the poor.

The Washington Post (1984 October 14). Close to home: Should we require shelter for the homeless?

The Washington Post (1985 February 5). Barry opens homeless aid drive: coalition seeks $150,000 in emergency funds.

The Washington Post (1986 May 21). Homeless initiative ruled valid.

The Washington Post (1990 November 4). Referendum 005: don't throw good money after bad.

Walters, Ronald, & Travis, Toni-Michelle (eds). (2010). *Democratic Destiny and The District of Columbia.* New York: Rowman and Littlefield.

WAMU 88.5 FM (2018 March 12) Homeless advocates question bowser's plan to close D. C. general shelter this year. https://wamu.org/story/18/03/14/homeless-advocates-question-bowsers-plan-close-d-c-general-shelter-year/

Ward, Kevin (2017). Policy mobilities, politics and place: The making of financial urban futures. *European urban and regional studies*, *25*(3), 266–283.

Ward, Kevin (2018). Urban redevelopment policies on the move: Rethinking the geographies of comparison, exchange and learning. *International Journal of Urban and Regional Research*, *42* (4), 666–683.

Webber, Sophie (2015). Mobile adaptation and sticky experiments: Circulating best practices and lessons learned in climate change adaptation. *Geographical Research*, *53*(1), 26–38.

Wells, Katie (2014). Policyfailing: The case of public property disposal in Washington, DC. *ACME: an International Journal for Critical Geographies*, *13*(3), 473–494.

Wells, Katie (2015). A housing crisis, a failed law, and a property conflict: The U.S. urban speculation tax. *Antipode*, *47*, 1043–1061.

Williams, Lois, Jackson, S., Trinity, F., & Schorr, S. (1993). The District of Columbia's response to homelessness: Depending on the kindness of strangers. *D.C. Law Review*, *2*(1), 47–90.

Urban policy (im)mobilities and refractory policy lessons: experimenting with the sustainability fix

Aida Nciri and Anthony Levenda

ABSTRACT

This paper bridges scholarship on policy mobilities and urban climate change experimentation to analyze the ways in which innovative low-carbon policies fail to diffuse. It argues that urban experiments become strategic learning tools that allow dominant actors in urban environmental politics to map pathways for a sustainability fix, test new low-carbon interventions, and gain knowledge of pathways for growth. Through a case-study of a solar district heating demonstration project in the Calgary metropolitan region, we suggest that these experiments allow powerful actors to mobilze "perverse policy lessons" in order to construct "policy failures" in cases that do not meet their requirements for a sustainability fix. Our analysis elucidates material and discursive strategies mobilised by dominant actors to selectively circulate knowledge that defines an urban experiment's success or failure. We highlight two takeaways for future scholarship on urban environmental governance and policy mobilities.

Introduction

This paper focuses on the Drake Landing Solar Community (DLSC), a solar district heating demonstration project, to understand the role of policy learning in the urban politics and governance of low-carbon transitions. The DLSC was completed in 2007 in the Town of Okotoks in the Calgary Metropolitan Region (CMR) of Alberta, Canada.[1] It was supported by Natural Resources Canada (NRCan), a department of the Canadian Federal Government, and leveraged partnerships with industry to demonstrate a model for low-carbon heating in Northern cities. Despite technical success, broad support, and intentions for widescale replication, the lessons gained in the DLSC have not been emulated in Alberta, nor has it spurred reproduction at a larger scale in Canada. Why would such a successful experiment for low-carbon urbanism fail to be emulated in communities throughout Canada as it was intended to do? Why didn't these policy lessons diffuse outside of the CMR? Our paper analyzes this inability to realize the "longer-term benefits" of replication and diffusion through the concepts of policy learning and the "sustainability fix". The sustainability fix refers to urban governance strategies that simultaneously meet profit-making and environmental goals without addressing structural inequalities. We suggest that "perverse policy lessons" are mobilized by powerful interests to construct "policy failures" in cases that do not meet their requirements for the sustainability fix. In this way, we see policy learning as relational and power-laden, not dualistic or apolitical.

The DLSC project was inspired from similar systems implemented in Scandinavian countries and was "imported" to Alberta. It was rolled out as a turnkey project by NRCan and a coalition of local actors. The location of this low-carbon project is surprising for two reasons. First, the province's strong economic reliance on oil and gas and weak policies on climate change forecloses many political pathways for low-carbon transition (Diamanti, 2016).[2] Alberta's petro-dominated economy and global position in fossil capitalism has made imagining and practicing low-carbon transitions particularly difficult (Adkin, 2016; Wilson, Carlson, & Szeman, 2017). Second, most Albertan municipalities who benefit from the growth and resource rents created by oil and gas industry are relatively passive in designing low-carbon urban policies. A survey conducted by the Albertan Municipal Climate Change Action Centre in 2014 indicates that only 40% of the respondents adopted a "formal or informal" community greenhouse gas emissions plan, about half of the respondents had no staff person responsible for climate change, and only one municipality had a full time designated person (Municipal Climate Change Action Centre, 2014). This highlights the important role of higher levels of government in urban low-carbon politics.

Within this context, we examine how the successful DLSC experiment failed to produce broader changes in urban socio-material systems. We also analyze how the policy lessons and technical knowledge it generated have been mobilised by the actors involved in the project *against* broader low-carbon transitions. We situate our case study within the growing literature on policy mobilities and urban (climate change) experimentation, contributing to the formulation of a conceptual framework that advances analyses of the politics of urban experimentation. Policy mobilities scholars highlight the importance of understanding how policy models, practices, and expertise are selectively mobilized, allowing different urban actors and institutions to intervene in urban development, propose specific solutions, and gather the political support and legitimacy needed to carry out particular interventions (Mccann, 2011; Temenos & Mccann, 2012a, 2013). Similarly, scholars studying low-carbon transitions and the urban governance of climate change have suggested these interventions are often rolled out in strategic projects, or "experiments," that test particular sociotechnical and policy solutions, generate practical lessons and technical knowledge, and rework how urban (carbon) governance is enacted (Bulkeley & Castán Broto, 2012, 2014). We bring these literatures into conversation to analyze the construction of urban experiments, such as the DLSC, and the mobilization of policy learnings. We show that policy failures are an important element of urban experimentation, but also suggest that policy lessons aid in the production of a sustainability fix (While, Jonas, & Gibbs, 2004). Our central argument is that politics of urban experiments are deeply intertwined with the social and power-laden construction of success and failure (and who defines them as such), and that actors can mobilize lessons from experimentation to slow down or block sustainable transition.

In the next section, we expand on our theoretical approach, building from literature on policy mobilities, urban experimentation, and urban political economy. We outline how policy failures (and sociotechnical experiments) are increasingly important to the reconfiguration of urban climate governance, yet the material interests of urban regimes for capital accumulation often limit the bounds of acceptable sociotechnical transformation and forms of experimentation while also capturing policy learnings to direct future strategies for growth. We then describe the case study of the DLSC in detail, providing empirical richness to our argument. We follow an extended case-study method as deployed by scholars of urban policy (Peck & Theodore, 2012). Our abductive analysis is based on a recursive "conversation" between empirical materials from our case and theory and concepts in urban geography (especially related to experimentation and policy mobilities). We utilize content and discourse analysis of interviews conducted with various actors involved in the project, as well as of development plans, policy at the federal and local level, and local media articles collected between 2013 and 2017. These materials were collected as a part of a five-year project on the politics of low-carbon transitions in Canada and France, specifically focused on district heating systems. Secondary source materials were gathered from the Town of Okotoks, local media, Natural Resources Canada, international climate organizations, and developer plans and websites. We supplemented the secondary source materials with half a dozen in-depth interviews with key stakeholders, and participant observation at community meetings and conferences in Calgary.

Our analysis shows how the DLSC was set up as a model for reproduction but failed to diffuse because of the economic interests of an existing urban regime who saw the model as a challenge to their bottom line interests in profitable property development and energy sales. The learnings from this experiment, were used to rework the regimes pathways for growth, ultimately undermining the low-carbon features, including the solar centralized heating system and green building models. We argue that the nature of urban experiments – as temporally limited, spatially bounded, and strategic interventions – provide opportunities for urban sociotechnical regimes to incorporate them into existing channels of growth. The power dynamics of sociotechnical regimes limits both the mobility of lessons and models generated/tested in urban experiments and the possibilities for more extensive low-carbon transition. We suggest that social and political production of successes and failures allows for experimentation with a "sustainability fix", providing the means through which regimes of developers, government, and utility interests can test-out different sociotechnical configurations for low-carbon transitions and experiment their potential to activate (or not) new channels of growth. We conclude with a discussion of how policy learning can be progressively understood to enable more just and reproducible low-carbon transitions, which entails an alternative urban politics of experimentation based in a pragmatic incrementalism.

Urban experimentation, policy learning, and the sustainability fix

In order to understand the case of DLSC in the broader context Alberta's low-carbon transition and how it illuminates broader trends in urban politics and governance, we build on three interrelated literatures. First, we draw on the literature on urban experiments to understand their role in reconfiguring urban governance and urban politics around learning and carbon reduction. Second, we build on the notion of the sustainability fix to understand the motivations driving these forms of urban environmental governance. Lastly, we draw on policy mobilities and policy learning to understand the ways in which policy lessons are selectively mobilized. Drawing these literatures together helps us understand the ways in which "refractory policy lessons" were generated in Okotoks, and the ways in which actors constructed the lessons to invalidate the efficacy of the DLSC model for replication or expansion. More broadly, our conceptual framework elucidates the politics of policy experimentation and mobility, especially related to green and sustainable urbanism, and thus is instructive for critical analyses of the successes and failures of urban sustainability policy.

The urban politics of experimentation

Bringing attention to the multitude of ways in which the urban intervenes in energy production, distribution, and consumption, Rutherford and Coutard (2013) argue cities have become key sites for the governance of energy transitions. Urban energy governance, as Rutherford and Jaglin (2015, p. 174) note, refers to the "multitude of ways in which urban actors engage with energy systems, flows, and infrastructures in order to meet particular collective goals and needs, framed or expressed in policymaking processes, but also in debates, contestations, and conflicts over policy orientations, resources, and outcomes." This definition suggests that urban energy governance does not only rest in official policy making procedures, but also in the various ways that urban actors interact with energy infrastructures. It also suggests that low-carbon politics are implicated in the reconfiguration of collective consumption, urban space, and provision of public services (Cohen, 2017; Jonas, Gibbs, & While, 2011).

This mirrors discussions of urban climate governance more generally, and the governance of low-carbon transitions (Bulkeley, 2010; Bulkeley, Castán Broto, & Edwards, 2014; Bulkeley, Castán Broto, Hodson, & Marvin, 2010). Bulkeley and Castán Broto (2012) document the multitude of low-carbon projects and policies in cities globally, referring to them as "climate change experiments" that are central to urban climate governance. These experiments "are purposively designed to trial the social and technical experience of responding to climate change, put new materials, technologies and social actions to the test, or develop knowledge within the city to respond to climate change" (Castán Broto & Bulkeley, 2013). Experiments, however, are not only opportunities for learning or developing best practices. As Bulkeley and Castán Broto note (2012, p. 368), "rather than regarding experimentation as an open-ended process orchestrated by the dynamics of learning the city, or outside of the proper governance of climate change in the city, [...] experiments are critical sites of urban climate politics."

Scholarship on urban experimentation has engaged with various conceptions of urban politics, usually relating to climate, environment, or sustainability-related concerns. Evans (2011, p. 233) argues that the socio-ecological systems approach to urban adaptation, for example, has set out to create the city as a place for experimentation, opening up political possibility: "The central role afforded to experimentation in current manifestations of urban sustainability undoubtedly offers up a potential space

for more playful or insurgent political engagements with urban infrastructure and
material form. If sustainability comes down to letting 1000 experimental flowers
bloom, then it matters who gets to experiment, and how." Similarly, the literature on
urban laboratories has identified the promises and pitfalls of experimentation (Evans &
Karvonen, 2014; Gopakumar, 2014; Karvonen & van Heur, 2014; Strebel & Jacobs,
2014). While experimentation may achieve new policy learnings or even radical changes
on a small scale, the questions around who benefits and who loses out in experimentation is determined on a case by case basis. If urban experimentation is centrally about
learning from real-world sociotechnical interventions, in a situated, change-oriented,
and contingent manner (Karvonen & van Heur, 2014), then we must ask: who defines
experiments, who controls the interventions, who shapes knowledge production and
circulation processes, and what happens when the experiment ends? In short, urban
experimentation has politics influenced by existing power relations in the city.

Recent scholarship on urban experimentation has grappled with these politics more
closely asking who "gets to take part" in experimentation at both the institutional and
practical levels (Evans, Karvonen, & Raven, 2016), especially considering that private
companies and firms are heavily involved in this mode of urban governance (Bulkeley
& Castán Broto, 2012; Evans & Karvonen, 2014). While experiments are often described
as central to generating and governing urban low-carbon or sustainability transitions
(Bulkeley et al., 2014; Moloney & Horne, 2015), the central question of power relations
has brought concern about contestation and resistance from dominant regimes or the
ability to manage such a widespread, power-laden process at all (Geels, 2014; Shove &
Walker, 2007). Evans et al. (2016) discuss three aspects of power relations important to
the politics of experimentation. First, power relations determining the *process* of urban
experimentation: who participated in designing and framing the experiment, who's
knowledge claims and discourse position is considered and how are socio-material
transformations selected and shaped (Bulkeley & Castán Broto, 2014). Second, they
argue the *outcomes* of experimentation are central to politics. How have urban experiments promoted a business-as-usual approach often resulting in the reproduction of
inequalities? Central to the discussion of policy failures, *how are successes defined and
measured*? Third, experiments can be a locus for radical change, a powerful political
strategy for grassroots movements to challenge and disrupt the status-quo. Urban
experiments can become "urban politics by other means" to reclaim the democratic
construction of the urban fabric and offer alternative socio-material pathways through
social innovations and new practices.

Growth politics and the low-carbon fix

Scholars investigating the politics of urban development have shown how environmental policy considerations are being mainstreamed to enhance urban competitiveness and
gain boarder political support (Jonas et al., 2011; Long, 2016; While et al., 2004; While,
Jonas, & Gibbs, 2010). While et al. (2004) advance the concept of a "sustainability fix"
building on Harvey's (Harvey, 1981) concept of the spatial fix to analyze the ways
discourses of sustainability are enrolled in an entrepreneurial approach of urban
growth. Caught between growing pressure to adopt local policies on environmental
issues on one hand, and inter-urban competition and state rescaling on the other, they

argue that cities are deploying green differentiation strategies to attract investment capital. Urban regimes selectively develop environmental policies with the aim to reproduce or create new channels of growth:

> The historically contingent notion of a 'sustainability fix' is intended to capture some of the governance dilemmas, compromises and opportunities created by the current era of state restructuring and ecological modernization. [...] sustainable development is itself interpreted as part of the search for a spatio-institutional fix to safeguard growth trajectories in the wake of industrial capitalism's long downturn, the global 'ecological crisis' and the rise of popular environmentalism. [...] The notion of sustainability fix does not deny progress on ecological issues, but draws attention to the selective incorporation of ecological goals in the greening of urban governance." (While et al., 2004, p. 551)

The concept of sustainability fix entails a reconfiguration of the relationship between private and public actors, urban policies, and the built environment, and "in this sense the 'sustainability fix', for both the public and private sector, is often as much about changes in political discourse as it is about material change in the ecological footprint of economic activity" (While et al., 2004). Expanding on the sustainability fix to analyze the broader relationship between ecological crisis and urban growth politics, Jonas et al. (Jonas et al., 2011) suggest carbon control is the central discourse, strategy and struggle around urban development in the "New Environmental Politics of Urban Development" (NEPUD). They explain that this entails a "retooling" of the urban growth regime, public-private partnerships, and urban fiscal incentives around carbon reductions throughout the urban fabric. Low-carbon transition here is thoroughly embedded not only in interurban competition as "the dominant mode of political calculation in urban governance", but also manifest in contestations between urban growth coalitions and a low-carbon polity in struggles around collective consumption, public services, public infrastructures, and environmental regulations that might conflict with or guide strategies of growth (Jonas et al., 2011, p. 2539). Thus, the NEPUD provides a conceptualization of the interface between policy flows and territorial interests that might influence policy mobilities (Chang, 2017; Temenos & Mccann, 2012a) and the shaping of urban experiments (Bulkeley & Castán Broto, 2014; Evans & Karvonen, 2014).

Recent scholarship on urban environmental governance has expanded on the sustainability fix and the NEPUD, arguing that "new urban environmental regimes" represent three dominant trends: (1) development of an economic sector of green urbanism characterized by neo-managerial instruments of control and new modalities of competitiveness; (2) increasingly mobile, (trans-)locally produced and transformed policies aimed at building green urbanist reputations; and (3) broad movements towards sustainability post-politics that recreates exclusionary urban politics (Rosol, Béal, & Mössner, 2017, p. 1711). These new urban environmental regimes are defined by six features: they are growth-oriented, neo-managerial, best-practice driven, socio-spatially selective, city-centric, and post-democratic (Rosol et al., 2017). Centrally, these regimes promote sustainability in a largely symbolic way, to present cities as green while reproducing the status quo. This is evidenced by the urban environmental policies focused on growth-oriented strategies that are predominantly state-led and driven by public actors in government agencies.

In order to understand how the politics of urban experimentation and the sustainability fix influence the circulation of policy models and knowledge, we turn to the scholarship on policy mobilities and policy learning. These literatures help us understand the dialectics of policy failure and success, and the role of experimentation in testing pathways that simultaneously assure capital accumulation and assuage environmental concerns. In our specific case, a publicly-funded low-carbon urban experiment provides a learning site for dominant growth coalitions to test channels of economic growth in "green" urban socio-technical configurations.

Power and policy learning

Policy mobilities literatures brings consideration of the role of knowledge circulation from the site of experiment to the "outside", and the constitution of networks that produce and circulate certain forms of knowledge (Mccann, 2011; Temenos & Mccann, 2013). The policy mobilities approach focuses on "following policies and 'studying through' the sites and situations of policy making" (McCann & Ward, 2012). Its methodological undertaking and inquiry (Cochrane & Ward, 2012; Peck & Theodore, 2012) critically investigates two intertwined processes in the construction and travel of urban policies. First, the *selective and power-laden process* of circulation of knowledge and policy through formation of transnational networks (or constellation) of actors. Scholars analyze why policy travels from place to place, how policies travel, and who makes policies mobile for whom and what (Mccann, 2008; Montero, 2017; Temenos & Mccann, 2012b; Wood, 2015a). Secondly, it focuses on how policies are constructed and how policies change as they circulate. Scholars examine how networks of actors transform and adopt elements of policy, as well as how place-specificities impact processes of mutation (Muller, 2014; Peck, 2012; Wood, 2015b).

These considerations are central to understanding the power relations behind the branding of "best practices" and the selective production and circulation of policy knowledge by different networks of actors (Baker & Temenos, 2015; Peck & Theodore, 2012; Temenos & Mccann, 2013). Critical attention to these power relations have pointed towards the importance of accounting for policy failures, and more centrally to deconstructing the dualisms of failure/success and immobility/mobility (Mccann & Ward, 2015). Similarly, Baker and Temenos (Baker & Temenos, 2015) argue that defining a policy as a success or failure are "power-laden" notions, resulting from the ways in which evaluations are socially and politically constructed.

An underlying assumption of the urban experimentation literature is that learnings and knowledge generated through a project might be mobilized by participants for replication and diffusion. Urban experiments aim to be replicated in "other locales" (Karvonen, Evans, & van Heur, 2014) or to be sustained and scaled up "to the entire, or at least a significant portion, of the urban landscape, through the transformation of urban planning practices" (Bulkeley & Castán Broto, 2014). This epistemology aims for "transformative learnings" – i.e. learning directed toward change and transformation – as opposed to learnings directed toward blockage and reproduction of the status-quo. Haughton and McManus (Haughton & Mcmanus, 2012) describe this as "perverse policy learning," building on Sites (2007) conceptualization of the ways in which cities become "social laboratories for the corporate and policy 'learning' that enhances neoliberal flexibility and governance."

Following Mccann and Ward (2015) emphasis on relational thinking against dualisms such as policy success and failure, we suggest this can also be applied to the concept of learning. Instead of positing that learning is positive or negative, we follow McFarlane's (2011, p. 362) conceptualization of learning as "crucial to how urbanism is produced and to how different constituencies respond to it." Learning in this regard is always potentially transformative, but also always power-laden. In support of powerful interests such as those involved in "creative" or "smart" city agendas, there is a tendency to focus on learning for knowledge-economies (Hollands, 2008; Krätke, 2012). Yet, as McFarlane (2011, p. 363) notes, this is "learning only in name and the purpose is to confirm what is already known, or to support existing politico-corporate agendas, while in other cases urban learning may be reduced to a direct or indirect process of imposition or instruction rather than dialogue and reflection."

Green urban experiments can thus be conceptualised as a strategy for dominant actors to test potential channels of growth associated with low-carbon transformations and to use these policy learnings for future growth-oriented agendas. In many cases, these urban experiments serve as a sustainability fix to simultaneous fuel growth and mitigate environmental impacts. If urban experiments are not profitable, they are often branded as a failure. Of course, in practice determining the success or failure of an urban experiment is more complex and power-laden. Constructing a narrative of failure or success centers around sorting out and selecting socio-technical interventions and policy models that are the most "promising" while putting aside those which are not profitable – or even threatening to economic interests. The construction of these narratives are tied to the broader development of a sustainability fix. This means analyses of urban experiments must take into account place-specific power relations and their role in the mobilization of policy learnings.

This scholarship on policy learning, urban environmental governance, and urban experimentation provides a conceptual framework for our analysis of the ways in which powerful actors selectively mobilize policy learnings associated with urban socio-material transformation. While urban experimentation offers opportunities for cities to progressively act on climate change or environmental issues, it also offers opportunities for business-as-usual approaches to construct a green image and test a particular sustainability fix. Our synergistic review of these literatures demonstrates that (1) urban experimentation is political and power-laden, and (2) policy success and failure is socially and politically constructed, coopting the radical potential of urban learning to serve the interests of profit-motivated developers and utilities. In the following sections, we show the impetus for experimentation in Okotoks was driven by the potential for the DLSC pilot project to generate economic opportunities associated with a new low-carbon district heating system. We identify perverse policy learning by analyzing the knowledge and practices that are purposefully captured and immobilized by urban environmental regimes to prevent radical changes that could threaten their dominant position in the economic status quo.

The drake landing solar community experiment

In order to provide background on Alberta's urban energy and environmental politics, this section contextualizes Okotoks in the broader Calgary metro-politics. We provide a brief overview of previous local-level energy and environmental planning and policy. Then we provide background information about the DLSC and its potential for low-carbon heating in Northern cities.

Petro-state urban politics

The economic boom of the oil and gas industry have impacted urbanization in Alberta. This has brought jobs and population growth to the Edmonton and Calgary metro areas and put new pressures on urban governments to develop sustainably. Within the Calgary metropolitan region (CMR), the Town of Okotoks population has increased dramatically over the last two decades, from 8,528 in 1996, 17,150 in 2006, and to 28,881 in 2016. Most of this growth has been absorbed through low-density developments and extension of municipal borders through successive land annexation. Single-detached houses represent 77% of the dwelling stock in the town (Canada Census 2006, 2016).

Okotoks is characteristic of growing suburbs in the Canadian "in-between" often experiencing trouble meeting infrastructural needs and service demands (Filion & Keil, 2017; Young & Keil, 2010). In the late 1990s, the growth capacity of Okotoks was threatened by provincial water regulation and water quota management. To authorize its growth strategy, the municipality had to secure a minimum amount of water per capita. Okotoks had to cap its population growth to 30,000 inhabitants in its 1998 Municipal Development Plan. The same year, the Town of Okotoks adopted a sustainability strategy plan, mostly focused on water conservation. After significant investment in water savings programs, and the negotiation of an agreement with the City of Calgary, the municipality managed to cope with water requirements and increased its access to new water resources. In 2012, the town reviewed its municipal development plan and decided to "transition[n] from the finite growth model of the 1998 Legacy Plan to one of continued managed growth" to accommodate an expected increase of in population to over 80,000 over the next three decades (Town of Okotoks, 2016).

In Okotoks, the real estate sector, home-builders and developers represent a significant share of the local economy, and play a central role in politics and planning. Municipal policy in Calgary in the late 1990s and 2000s was largely focused on implementing new urbanist concepts and sustainability goals combined in a "smart growth" agenda (Grant, 2009). This strategy included a "sustainable suburbs" policy that was slow to be implemented in the late 2000s because of a push by developers against regulation. In periods of high growth during the tar sands boom, developers appeared to be more willing to support plan and policy changes for smart growth, such as dense and transit-oriented development, especially in privatized suburban developments (Grant, 2009). In this push for working towards more sustainable suburbs, Okotoks had been positioned as a site for testing solar district heating technology in newly built residential developments.

Assembling the DLSC

The Drake Landing Solar Community (DLSC) is an innovative energy district heating system built in Okotoks in 2007. District heating systems (DHS) are decentralised heating infrastructure that supplies heat from one or several thermal plants to many buildings through an underground network of pipes. Heating plants can use various sources of energy: fossil fuels, low-carbon energy sources including waste heat recovery, biomass, geothermal and thermal solar. The flexibility of energy sources and the possibility to use renewable energy makes district heating a widely applicable tool for reducing greenhouse gas emissions in cities.

While DHS have been widely used in Europe and are often cited as a key sociotechnical intervention for urban low-carbon transitions with its own inherent organizational and implementation concerns (Gabillet, 2015; Hawkey, Webb, & Winskel, 2013; Webb, 2015), this infrastructure has not been widely implemented in residential applications in Canada (CIEEDAC, 2014). Over the past two decades, there has been a growing interest in DHS across Canadian cities. This interest has been bolstered by European experience and the advocacy of municipal network associations, such as the Federation of Canadian Municipalities. Even at the provincial level, there have only been three DHS pilots. The DLSC stands out because it showcases an innovative low-carbon heating technology, as opposed to natural gas; it was initiated by an individual working for a federal agency; and the pilot was built in a low-density suburban development.

The DLSC showcases a seasonal solar thermal storage technology (see Figure 1) The system connects 52 energy-efficient detached, single-family homes. Each dwelling has solar thermal arrays installed on the garage roof. Heat is captured and stored in the ground through an underground system of pipes and reservoirs during the summer. During winter, heat stored in the ground is released to heat the homes. The system avoids the emissions of 5 tonnes of carbon per house per year – including two tonnes due to the high energy performance of the houses.

The DLSC project was initiated and managed by an engineer of a federal agency within the Canadian Ministry of Energy. The project manager's career in solar energy had given them experience and knowledge on solar thermal projects implemented in Scandinavian countries and Germany. In the late 1990s, "all the factors sort of laid down to help [the project] come about" (Project Manager, Interview, January 2015). This included growing concerns over climate change, increasing energy prices, and federal funding allocated to renewable heating projects in the residential sector.

With these drivers for implementation motivating the project, the main challenge was to find a place to anchor the demonstration project and to "assemble a team" consisting of a municipality, a developer, a home builder, and a utility. Through a process of networking amongst the project team, a low-density residential development was identified in Okotoks. The project was approved because no homeowner or project team member would pay any additional cost for the solar DHS. As the project manager explained:

> We assembled that team. But the pitch that I made in order to bring these people together and agree to the proposal, was that we will find the funding to pay for the solar technologies and the storage technologies [...] That's what brought them in. They saw they didn't have to come up with extra cash in order to pull this off. That they would rely

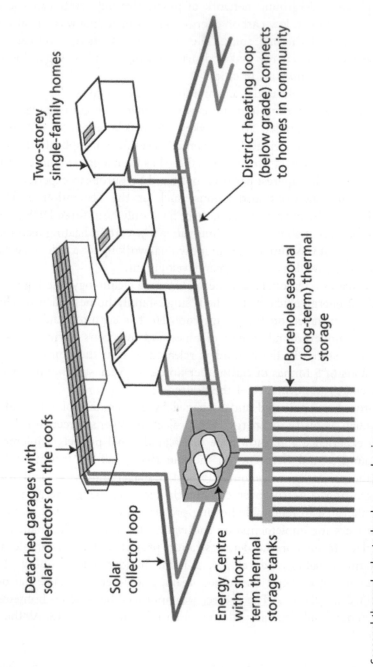

Figure 1. Seasonal thermal solar technology mechanisms.

Source: (Canmet Energy, 2009) Url: https://www.nrcan.gc.ca/sites/www.nrcan.gc.ca/files/canmetenergy/files/pubs/DrakesLanding(ENG).pdf
No need for permission request. See: https://www.canada.ca/en/transparency/terms.html

on us to sort the funding for that. [...] This is how we get them in initially, by removing this financial risk from them. (Project Manager, Interview, January 2015)

Beyond the local approval for this pilot project, technical expertise was required for modelling and designing the thermal storage system. A team of international experts was recruited to design the system in which they deployed a Swedish company's model that was successful in earlier energy storage projects. Construction started in 2005 and the project was completed in 2007.

Demonstrating solar district heating

The total cost of DLSC was CAD$7 million or approximately $134,000 per house. This cost included energy efficiency upgrades, but excluded the cost of land and homes construction which was the responsibility of the developer and homeowners. The "experimental" portion of the project was almost entirely funded by two major federal grants (a $2 million grant from the federal Technology Early Action Measures fund and a $2.5 million grant from the Federation of Canadian Municipalities). The provincial government also awarded the project a small grant.

According to the main page of the project website, the objectives of this urban pilot were twofold: to demonstrate the feasibility of seasonal solar thermal storage in Canada's cold climate; and to be replicated at a larger scale, in other Canadian cities (Drake Landing Solar Community, n.d.). From the start of operation, the DLSC has successfully demonstrated the technical feasibility of seasonal thermal storage technology in a cold climate. Over the past six years, the system provided over 90% of space heating needs, surpassing initial expectations. This success was recognized by several prestigious national and international energy prizes and awards, such as the Golden Energy Globe World Award from the Energy Globe Foundation in 2011, a first for a Canadian project. In 2013, the project won the International Energy Agency (IEA) Solar Heating and Cooling (SHC) Programme Solar Award, which "recognises not only the excellent results of the project, but also the pioneering spirit of the involved partners" (IEA Solar Heating Programme, 2013). Building on this publicity, in 2012, a Korean delegation visited the DLSC to learn from the project and to see if it could be replicated in developments for the 2018 Winter Olympics.

Despite this broad success and notoriety, ten years after its completion, the DLSC remains a North American first that failed to be replicated. Discussions were initiated with the municipality of Fort McMurray (Alberta) and a few municipalities in Nova Scotia, and feasibility studies were conducted in Whitehorse (Yukon) as well as in Okotoks for an additional 1000-unit system (Rodham, 2011). Yet, none of the municipalities followed up, mostly because of the high upfront costs of the system, and the absence of public funding available to replicate the project. The next section analyzes the role of power relations in the construction of this policy failure in the DLSC.

Solar trials and economic tribulations: constructing policy failure

To understand how policy failures are constructed and how they influence urban environmental governance, we unpack the politics of experimentation through an analysis of the DLSC. We discuss three moments in the production and mobilization of policy learnings: (1) sociotechnical construction of the urban experiment, (2) shaping of policy failure, and (3) blockage of radical change.

Sociotechnical construction

The DLSC was inspired by seasonal solar thermal storage system developed in Scandinavian countries. Yet, when implemented in Okotoks, the pilot project underwent a series of significant socio-technical alterations that weakened its economic performance. Two key spatial features of district heating systems (DHS) were neglected during the transfer of knowledge: autonomy and self-sufficiency on one hand, and density on the other. DHS lays at the intersection between urban planning and energy systems, and like many infrastructure systems, they are place-specific and place-sensitive. Given DHS are capital-intensive infrastructures and rely on economies of scale to operate efficiently, their economic viability depends on the *energy density* of an area, i.e. heating load per kilometre of pipe network. The higher the building density, the more secure the heat load, and the easier to secure a reasonable return on investment.

In the case of DLSC, the system was poorly designed in regard to the principles of economy of scale. The system was designed and built in a low-density suburban development. Construction was made possible by the provision of public funding covering the upfront capital costs. The revenues generated by the fifty-two single-detached houses are barely sufficient to cover operations and maintenance costs, and there are not enough reserve funds to pay for major repair or replacement. In addition, the costs of the system were overrun due to infrastructural redundancy: the fifty-two homes are also connected to the conventional natural gas distribution system that provides for the remainder of space heating and hot water needs. This led to higher energy bills for connected households who pay fees for the DHS and a significant annual connection fee for the natural gas distribution system.

By contrast, energy efficient homes were not an initial component of the demonstration project. The official news release issued by (NRCAN, 2005) does not mention energy efficiency measures. Energy efficiency became a central feature of the DLSC only when the modeling phase underscored the prohibitive costs of designing a system that would meet the heating needs of conventional homes. Based on the learnings from modeling, the project manager decided to make the project energy efficient. The only official thermal insulation standard at that time was the R-2000 certification. This certification developed by NRCAN required homes to consume thirty percent less energy than conventionally built homes, that was the equivalent of a reduction of energy consumption by thirty-seven Gigajoules per year.

The highly insulated homes decreased the necessary storage capacity of the system by forty-five percent. Energy savings also contributed to reduced carbon emissions by two tonnes per house per year. This is compared with the 5.6 tonnes avoided by the entire system (energy efficiency and solar storage). Energy efficient homes are a low-cost intervention compared to the rest of the system, at only $6,400 (Howell-Mayhew Engineering, 2008). Energy efficiency contributed to thirty-five percent of the total carbon savings while representing only five percent of the initial capital costs of the project.

Energy efficient homes were an additional part of the urban experiment's sociotechnical construction. The R-2000 performance helped the project's technical success in the face of limited financial resources of the project by enabling the system designers to decrease the size, and the corresponding cost, of the storage system. As one engineering consultant noted, the experiment did not consider energy efficiency as a central feature, but focused on the demonstration of the solar thermal heating technology:

> In fact (...) I said to them [the other engineering consultants]: "Well, you are going to make it R-2000 energy efficient first, aren't you?" And he looked at me as if I was from Mars. He thought I was absolutely crazy to make a project that was energy efficient and he said: "No, no, we just want to make it solar". (**Engineering Consultant, Interview 22 October 2015**)

The sociotechnical construction of the DLSC was focused on "proving" the concept of solar thermal district heating. While common sense engineering knowledge would always recommend efficiency first, this was developed as an add-on to the DLSC's demonstration. Corresponding to much of the critical scholarship on urban experimentation (Halpern, LeCavalier, Calvillo, & Pietsch, 2013; McLean, Bulkeley, & Crang, 2015), the focus on demonstration and technological testing often blinds opportunities for more radical (and perhaps simpler and cheaper) solutions for responding to climate change. In this case, the DLSC's tunnel-vision view for implementing solar DHS nearly was only made possible by integrating basic forms of carbon reduction technology, well-insulated homes.

(Perverse) learnings: constructing failure

The DLSC experiment produced four policy lessons that were mobilized selectively in order to showcase the sustainability signature of Canada's first solar thermal DHS while also undermining its practical and economic viability:

(1) Confirmation of the potential for a technological fix to reduce greenhouse gas emissions;
(2) Demonstration of the exorbitant costs of the technological fix compared to its environmental benefits, in a province with low energy price and weak carbon policy;
(3) Demonstration of financial unsustainability of the operation and maintenance of a DHS built in a low-density development;
(4) Demonstration of cost-efficiency and carbon-efficiency of energy efficiency homes compared to the thermal solar technology.

These learnings are based on financial and material characteristics of the project. The mobilization and translation of these learnings by the participants vary in that the first two were expected by the participants but the second two were not. Correspondingly, the socio-material implications of these learnings are such that some aligned with developer and utility interests and others challenged them.

The selective ambitions to focus on a technological demonstration and not on the infrastructural aspect of DHS were well-understood before the implementation of the project by the home builder, the developer, the utility, and the Town of Okotoks. During the conception phase, the project manager organized a study tour in Sweden, the Netherlands, and Germany to educate the participants on seasonal solar thermal storage systems to "bring [them] up to speed" (Interview, DLSC Project Manager, January 2015). The participants had the opportunity to connect with stakeholders involved in the construction and operation of those systems. The project manager described the aim of the project:

> To be clear with what we were doing with the project, this project – the size of it being only 52 homes – was big enough to demonstrate the concept of seasonal storage [...] but it is not the size in order to move towards commercialisation. You will really never build a project this size for it to be commercially viable. You have to go to a much, much larger project. (Interview, DLSC Project Manager, January 2015)

Study tours play an important role in policy knowledge production and exchange (Cook, n.d.; Cook, Ward, & Ward, 2015). In the DLSC, this enabled local actors to be aware of mutations from best practices" as demonstrated in Europe (e.g. technological focus and neglect of DHS, low-density development, infrastructural redundancy). They also knew the outcome of the project – an uneconomic technical demonstration. In other words, during the design and implementation of the project, dominant actors knew that the DLSC differed from the 'inspirational" European models in two ways. First, they understood that the low-density suburban development and associated urban planning practices would be left unchanged. Second, they learned that, given the cost of the demonstration, the intention was not to replace fossil-fuel based heating systems with solar energy in the short- or medium-term.

These "predetermined" lessons were central to manufacturing a policy failure. Actors engaged in the project to benefit from the environmental branding and sustainability discourse while at the same time reproducing existing low-density urban planning practices and demonstrating non-threatening technological systems with zero financial risk. The limited scope of the project would be a way to continue business-as-usual development and assuage environmental concerns (at least in the short-term).

By contrast, the addition of energy efficiency measures and long term maintenance of the system has produced a set of unexpected practical policy lessons. First, it highlighted the relative cost-efficiency of high energy efficiency standards in reducing greenhouse gas emissions in comparison to the solar storage system. Second, it underscored the financial infrastructural burden resulting from urban sprawl and low-density development. Both of these lessons had the potential to challenge existing environmental regimes, but were actively quieted in discourse on the DLSC experiment.

At that time of the DLSC experiment, the province of Alberta had one of the least efficient building code standards in Canada despite the harsh winters (Alberta Energy Efficiency Alliance, 2009). In the context of rapid population growth, energy efficient homes threatened home builders, developers, and energy utilities material interests, increasing the per unit costs and decreasing profits. In addition, energy efficient homes require industry support that includes worker training, new building practices, and increased construction costs.

> The builder did not want to do anything what so ever out of the ordinary, even taking their homes from their base level of energy efficiency to R-2000. They did not want to do it. The mindset back then was just so stupid against anything new [...] R-2000 at the time (2004–2005) was amazingly ambitious, it was a leading edge. But there was a lot of resistant from home builders. (Engineering Consultant #1, Interview 22 October 2015)

For natural gas utilities, higher energy efficiency standards would have translated into lower energy sales, and thereby a lower turnover. In addition, the growing cost of sprawl on municipal budget, especially the cost of operation and maintenance of urban infrastructures, started to challenge low-density residential development and land annexation. For example, the City of Calgary projected that "combined cumulative operating and capital revenue and funding deficiency ... to be approximately $7 billion over the next 10 years" (City of Calgary, 2011). To resorb this deficit the City reviewed its Municipal Development Plan in 2009 to limit greenfield development, with the "endeavour to accommodate 50% of Calgary's future population growth over the next 60–70 years within Developed Areas of the city" (City of Calgary, 2009).

Blockage of radical change?

Developers and utilities involved in the DLSC mobilized two strategies to shape policy learnings: selective circulation of policy knowledge and reproduction of a market solution narrative. The first strategy built upon the tension between the technological success of the experiment and its economic failure. On one hand, the awards and accolades received by the DLSC project from established organizations in the energy landscape provided credentials for the solar DHS. This credential permitted the technology to disrupt the dominant urban energy regime in principle, yet, the exorbitant costs for infrastructure and the low price of natural gas foreclosed pathways for market penetration in practice. This situation is well summarized by the home builder: "We've proved it can be done. Now we just have to prove it makes economic sense" (Rodham, 2011). By making visible the technological success of the project and its economic failure, the dominant actors were able to use the discourse of ecological modernization to suggest the DLSC was a failure and would only be feasible if the costs of solar heating technologies decrease and the scale of the systems could be increased. Second, they limited the focus on low-carbon interventions embedded in the project. The focus on seasonal thermal solar storage and the ecological modernization narrative diverted attention from energy efficiency features and the lessons for low-carbon urban form.

The discourse around the replication of the project has focused on the solar thermal heating technology, not about the more "common sense" lessons regarding energy efficient homes. As a result of this selective discussion of the experiment's success, the focus is on the non-threatening technology rather than on low-tech energy saving know-how. In addition, the discursive framing of the DLSC experiment does not challenge the unsustainable patterns of low-density suburban development in Alberta. As described by Sibbitt et al. (2015, p. 36), the project was built to conform to existing patterns of development: "At first glance, the tidy two-story houses lining two streets in a Canadian suburb look much like thousands of other homes that surround them; but a district heating system that stores summer's abundant solar energy to heat the homes during winter makes this community a global pioneer in heat storage technologies." This diverts the attention from the intrinsic economic and environmental unsustainability of greenfield, low-density and car-oriented development, and displaces alternative urban planning practices based on compact, mixed-used, and transit-oriented developments. The economic narrative disqualifying the solar technology because of its high cost also eliminates broader consideration of the political construction of energy prices, and the possibility for the Provincial Government to influence energy prices by introducing regulations effectively pricing carbon. In the context of Alberta's petro-politics, such radical changes would force consideration of alternative sociotechnical configurations.

To summarize, three policy lessons and their associated urban socio-material implications were marginalised by the coalition of local actors and government institutions involved in the DLSC project: energy efficient homes, compact development, and urban energy systems with higher fossil fuel prices. The impact was a "showcase" project that earned sustainability credentials for low-carbon experimentation, but undermined such pathways and reiterated the dominant position of a fossil-fuel based energy regime.

Conclusion: mobilizing against transition

In this paper, we have argued that the selective mobilization of policy knowledge is central to the constitution of policy failure. Urban experimentation offers the potential for transformative learning and action, yet it is often deployed as a way to further the sustainability fix (Jonas et al., 2011; While et al., 2004). We suggested that "refractory" policy lessons are used to undermine low-carbon transitions, and that this is a growing feature of urban environmental governance not wholly captured by the idea of the sustainability fix, urban experimentation, or policy mobilities individually. Instead, we advocate for bringing these literatures together to analyze how dualisms of policy success/failure are weilded in support of, or against, urban low-carbon transitions. The case of the DLSC shows how analyses of urban environmental governance should be sensitive to policy learning and experimentation, but also to local, growth-oriented urban politics.

We highlight two key takeaways for future scholarship on urban environmental governance based on our analysis. First, analyses of the sustainability fix, or more broadly urban environmental politics, is increasingly tied to the global circulation of policy knowledge and to the broader trend of urban experimentation. Our case study suggests that urban experimentation can be part of a strategy for dominant regimes to test new channels of growth associated with ostensibly "green" sociotechnical interventions. Experiments offer one-off, contained projects, or what many transitions scholars call technological niches that offer opportunities for learning (Hess, 2016; Hommels, Peters, & Bijker, 2007). In new urban environmental regimes (Rosol et al., 2017), urban sustainability is often wielded to further the post-political, techno-managerial approaches to urban environmental governance. Local growth coalitions with convergent material interests are able to continue business as usual, with a flair of green credentials. In the case of DLSC, the energy utility, the developer, the home builder, and the municipality of Okotoks, had aligned material interests secured by a specific socio-material configuration entangling low-density residential development and a fossil-fuel based energy system.

Secondly, we argue for a need to go beyond identifying "perverse learning" that only bolsters existing, unsustainable development. Countering existing regimes means not only analyzing these strategies but also charting alternatives. Our analysis is instructive in advocating for ways in which policy learning might be progressively understood to enable more just and reproducible low-carbon transitions, which entails an alternative urban politics of experimentation. To realize this, there is a pressing need to establish agendas for experiments that directly challenge urbanization as usual. Experiments are not simply about the pursuit of novelty but about fundamentally altering the way that urban development is done in a particular place. The transformative potential of experimentation does not lie in a series of one-off experiments where knowledge gleaned is fed into existing policy mechanisms, but in challenging the status quo and offering alternatives by re-orienting policy and planning around inclusive innovation and learning activities. If we begin to understand experiments in cities as urban politics by another means (Evans and Karvonen, 2014), then the challenge of experimentation is to go beyond the existing constellation of actors and develop more participatory agendas that can imagine significantly different urban futures.

Notes

1. While Okotoks is not officially included in the Calgary census metropolititan area (CMA), it is a suburb of Calgary and located within the Calgary Metropolitan Region (CMR) and participates in the newly formed Calgary Metropolitan Region Board (CMRB).
2. In Canada, provinces and territories hold large competences and responsibilities related to energy and climate change.

Disclosure statement

No potential conflict of interest was reported by the authors.

References

Adkin, Laurie. (2016). *First world petro-politics: The political ecology and governance of Alberta.* Toronto: University of Toronto Press.

Alberta Energy Efficiency Alliance. (2009). *Energy efficiency in the provincial building code.* Calgary.

Baker, Tom, & Temenos, Cristina. (2015). Urban policy mobilities research: Introduction to a debate. *International Journal of Urban and Regional Research, 39*(4), 824–827.

Bulkeley, Harriet. (2010). Cities and the Governing of Climate Change. *Annual Review of Environment and Resources, 35*(1), 229–253.

Bulkeley, Harriet, & Castán Broto, Vanesa. (2012). Government by experiment? Global cities and the governing of climate change. *Transactions of the Institute of British Geographers, 37,* 1–15.

Bulkeley, Harriet, & Castán Broto, Vanesa. (2014). Urban experiments and climate change: Securing zero carbon development in Bangalore. *Contemporary Social Science, 9*(4), 393–414.

Bulkeley, Harriet, Castán Broto, Vanesa, & Edwards, Gareth A.S. (2014). *An urban politics of climate change: Experimentation and the governing of socio-technical transitions.* London: Routledge.

Bulkeley, Harriet, Castán Broto, Vanesa, Hodson, Mike, & Marvin, Simon. (2010). *Cities and low carbon transitions.* London: Routledge.

Canmet Energy. (2009). Case study: Drake Landing Solar Community (DLSC) Okotoks, AB. *Leadership in EcoInnovation,* (Spring), 4.

Castán Broto, Vanesa, & Bulkeley, Harriet. (2013). Maintaining climate change experiments: Urban political ecology and the everyday reconfiguration of urban infrastructure. *International Journal of Urban and Regional Research*, *37*(6), 1934–1948.

Chang, I. Chun Catherine. (2017). Failure matters: Reassembling eco-urbanism in a globalizing China. *Environment and Planning A*, *49*(8), 1719–1742.

CIEEDAC. (2014). *District energy inventory for Canada, 2013*. Burnaby, BC: Simon Fraser University.

City of Calgary. (2009). *Calgary's municipal development plan*. Calgary: Author.

City of Calgary. (2011). *Long range financial plan*. Calgary: Author.

Cochrane, Allan, & Ward, Kevin. (2012). Researching the geographies of policy mobility: Confronting the methodological challenges. *Environment and Planning A*, *44*(1), 5–12.

Cohen, Daniel Aldana. (2017). The other low-carbon protagonists: Poor people's movements and climate politics in São Paulo. In M. Greenberg & P. Lewis (Eds.), *The city is the factory: New solidarities and spatial strategies in an urban age* (pp. 140–157). Ithaca, NY: Cornell University Press.

Cook, Ian R. (n.d.). Showcasing Vällingby to the world: Post-war suburban development, informational infrastructures and the extrospective city. *Planning Perspectives*.

Cook, Ian R., Ward, Stephen V, & Ward, Kevin. (2015). Post-war planning and policy tourism: The international study tours of the town and country planning association 1947–1961. *Planning Theory and Practice*, *16*(2), 184–205.

Diamanti, Jeff. (2016). Transition in a petro province? The Alberta NDP in office. *Socialism and Democracy*, *30*(2), 187–202.

Drake Landing Solar Community. (n.d.). About DLSC. Retrieved from https://www.dlsc.ca/about.htm

Evans, J.P. (2011). Resilience, ecology and adaptation in the experimental city. *Transactions of the Institute of British Geographers*, *36*(2), 223–237.

Evans, J.P., Karvonen, Andrew, & Raven, Rob. (2016). The experimental city: New modes and prospects of urban transformation. In J. P. Evans, Andrew Karvonen, & Rob Raven (Eds.), *The experimental city* (pp. 1–12). London: Routledge.

Evans, J.P., & Karvonen, A. (2014). "Give me a laboratory and I will lower your carbon footprint!" – Urban laboratories and the governance of low-carbon futures. *International Journal of Urban and Regional Research*, *38*(2), 413–430.

Filion, Pierre, & Keil, Roger. (2017). Contested infrastructures: Tension, inequity and innovation in the global suburb. *Urban Policy and Research*, *35*(1), 7–19.

Gabillet, Pauline. (2015). Energy supply and urban planning projects: Analysing tensions around district heating provision in a French eco-district. *Energy Policy*, *78*, 189–197.

Geels, Frank W. (2014). Regime resistance against low-carbon transitions: Introducing politics and power into the multi-level perspective. *Theory, Culture and Society*, *31*(5), 21–40.

Gopakumar, Govind. (2014). Experiments and counter-experiments in the urban laboratory of water- Supply partnerships in India. *International Journal of Urban and Regional Research*, *38*(2), 393–412.

Grant, Jill L. (2009). Theory and practice in planning the suburbs: Challenges to implementing new urbanism, smart growth, and sustainability principles. *Planning Theory and Practice*, *10*(1), 11–33.

Halpern, Orit, LeCavalier, Jesse, Calvillo, Nerea, & Pietsch, Wolfgang. (2013). Test-bed urbanism. *Public Culture*, *25*(270), 272–306.

Harvey, David. (1981). The spatial fix – Hegel, Von Thunen, and Marx. *Antipode*, *13*(3), 1–12.

Haughton, Graham, & Mcmanus, Phil. (2012). Neoliberal experiments with urban infrastructure: The cross city Tunnel, Sydney. *International Journal of Urban and Regional Research*, *36*(1), 90–105.

Hawkey, David, Webb, Janette, & Winskel, Mark. (2013). Organisation and governance of urban energy systems: District heating and cooling in the UK. *Journal of Cleaner Production*, *50*, 22–31.

Hess, David J. (2016). The politics of niche-regime conflicts: Distributed solar energy in the United States. *Environmental Innovation and Societal Transitions, 19*, 42–50.

Hollands, Robert G. (2008). Will the real smart city please stand up? *City, 12*(3), 303–320.

Hommels, Anique, Peters, Peter, & Bijker, Wiebe E. (2007). Techno therapy or nurtured niches? Technology studies and the evaluation of radical innovations. *Research Policy, 36*(7), 1088–1099.

Howell-Mayhew Engineering. (2008, July 21). *Borehole field in the the Drake landing solar community okotoks.*

IEA Solar Heating Programme. (2013). SHC SOLAR AWARD 2013 to Drake landing company – 52 homes heated with 98% solar. Retrieved from https://www.iea-shc.org/article?NewsID=48

Jonas, AE, Gibbs, D., & While, A. (2011). The new urban politics as a politics of carbon control. *Urban Studies, 48*(12), 2537–2554.

Karvonen, Andrew, Evans, J.P., & van Heur, Bas. (2014). The politics of urban experiments: Radical change or business as usual?. In Simon Marvin & Mike Hodson (Eds.), *After sustainable cities* (pp. 104–115). London: Routledge.

Karvonen, Andrew, & van Heur, Bas. (2014). Urban laboratories: Experiments in reworking cities. *International Journal of Urban and Regional Research, 38*(2), 379–392.

Krätke, Stefan. (2012). *The creative capital of cities: Interactive knowledge creation and the urbanization economies of innovation.* Oxford: John Wiley & Sons.

Long, Joshua. (2016). Constructing the narrative of the sustainability fix: Sustainability, social justice and representation in Austin, TX. *Urban Studies, 53*(1), 149–172.

Mccann, Eugene. (2008). Expertise, truth, and urban policy mobilities: Global circuits of knowledge in the development of Vancouver, Canada's 'four Pillar' drug strategy. *Environment and Planning A, 40*(4), 885–904.

Mccann, Eugene. (2011). Urban policy mobilities and global circuits of knowledge: Toward a research agenda. *Annals of the Association of American Geographers, 101*(1), 107–130.

Mccann, Eugene, & Ward, Kevin. (2015). Thinking through dualisms in urban policy mobilities. *International Journal of Urban and Regional Research, 39*(4), 828–830.

McCann, Eugene, & Ward, Kevin. (2012). Assembling urbanism: Following policies and 'studying through' the sites and situations of policy making. *Environment and Planning A, 44*(1), 42–51.

McFarlane, Colin. (2011). The city as a machine for learning: Translocalism, assemblage and knowledge. *Transactions of the Institute of British Geographers, 36*(3), 360–376.

McLean, Anthony, Bulkeley, Harriet, & Crang, Mike. (2015). Negotiating the urban smart grid: Socio-technical experimentation in the city of Austin. *Urban Studies, 53*(15), 3246–3263.

Moloney, Susie, & Horne, Ralph. (2015). Low carbon urban transitioning: from local experimentation to urban transformation? *Sustainability, 7*(3), 2437–2453.

Montero, Sergio. (2017). Study tours and inter-city policy learning: Mobilizing Bogotá's transportation policies in Guadalajara. *Environment and Planning A, 49*(2), 332–350.

Muller, M. (2014). (Im-)Mobile policies: Why sustainability went wrong in the 2014 olympics in Sochi. *European Urban and Regional Studies, 22*(2), 191–209.

Municipal Climate Change Action Centre. (2014). *Municipalities and climate change.* Government Institution.

NRCAN. (2005, March 30). Alberta solar-heating project first in North America. *News release.*

Peck, Jamie. (2012). *Constructions of neoliberal reason.* Oxford: Oxford University Press; Reprint edition.

Peck, Jamie, & Theodore, Nik. (2012). Follow the policy: A distended case approach. *Environment and Planning A, 44*(1), 21–30.

Rodham, Jason (2011). Living green in Drake landing. *Backbone*, pp. 22–24.

Rosol, Marit, Béal, Vincent, & Mössner, Samuel. (2017). Greenest cities? The (post-)politics of new urban environmental regimes. *Environment and Planning A, 49*(8), 1710–1718.

Rutherford, J, & Jaglin, S. (2015). Introduction to the special issue – Urban energy governance: Local actions, capacities and politics. *Energy Policy, 78*, 173–178.

Rutherford, Jonathan, & Coutard, Olivier. (2013, September). Urban energy transitions: Places, processes and politics of socio-technical change. *Urban Studies, 51*(7), 1353–1377.

Shove, Elizabeth, & Walker, Gordon. (2007). CAUTION! Transitions ahead: Politics, practice, and sustainable transition management. *Environment and Planning A, 39*(4), 763–770.

Sibbit, B, McClenahan, D, Djebbar, R, & Paget, K. (2015). Grounbreaking solar. *Igh Performing Buildings, Summer,* 36–46.

Sites, William. (2007). Contesting the neoliberal city? Theories of neoliberalism and urban strategies of contention. In Helga Leitner, Eric Sheppard, & Jamie Peck (Eds.), *Contesting neoliberalism: Urban Frontiers* (pp. 116–138). New York, NY: Guilford Press.

Strebel, Ignaz, & Jacobs, Jane M. (2014). Houses of experiment: Modern housing and the will to laboratorization. *International Journal of Urban and Regional Research, 38*(2), 450–470.

Temenos, Cristina, & Mccann, Eugene. (2012a). The local politics of policy mobility: Learning, persuasion, and the production of a municipal sustainability fix. *Environment and Planning A, 44*(6), 1389–1406.

Temenos, Cristina, & Mccann, Eugene. (2012b). The local politics of policy mobility: Learning, persuasion, and the production of a municipal sustainability fix. *Environment and Planning A, 44*(6), 1389–1406.

Temenos, Cristina, & Mccann, Eugene. (2013). Geographies of policy mobilities. *Geography Compass, 7*(5), 344–357.

Town of Okotoks. (2016). *Okotoks community sustainability plan 2016–2019.* Author.

Webb, Janette. (2015). Improvising innovation in UK urban district heating: The convergence of social and environmental agendas in Aberdeen. *Energy Policy, 78,* 265–272.

While, Aidan, Jonas, Andrew E.G., & Gibbs, David. (2010). From sustainable development to carbon control: Eco-state restructuring and the politics of urban and regional development. *Transactions of the Institute of British Geographers, 35*(1), 76–93.

While, Aidan, Jonas, Andrew, & Gibbs, David. (2004). The environment and the entrepreneurial city: Searching for the urban ` sustainability fix ' in Manchester and leeds. *International Journal of Urban and Regional Research, 28*(3), 549–569.

Wilson, Sheena, Carlson, Adam, & Szeman, Imre. (2017). *Petrocultures: Oil, politics, culture.* Montreal: McGill-Queen's Press - MQUP.

Wood, Astrid. (2015a). Multiple temporalities of policy circulation: Gradual, repetitive and delayed processes of BRT adoption in South African cities. *International Journal of Urban and Regional Research, 39*(3), 568–580.

Wood, Astrid. (2015b). The politics of policy circulation: Unpacking the relationship between South African and South American cities in the adoption of bus rapid transit. *Antipode, 47*(4), 1062–1079.

Young, Douglas, & Keil, Roger. (2010). Reconnecting the disconnected: The politics of infra-structure in the in-between city. *Cities, 27*(2), 87–95.

Beyond failure: the generative effects of unsuccessful proposals for Supervised Drug Consumption Sites (SCS) in Melbourne, Australia

Tom Baker and Eugene McCann

ABSTRACT
Focusing on the 20-year history of unsuccessful proposals for Supervised Drug Consumption Sites in Melbourne, Australia, this paper highlights the generative effects of apparent "failure" in policy-making and policy mobilization. Rather than framing thwarted proposals as categorical failures, we show how they altered parameters of policy acceptability, invigorated policy and practitioner networks, facilitated the development of allied programs, and, recently, inspired a successful SCS proposal. The paper argues that apparent policy failure and the potential for policy change must be evaluated and conceptualised in terms of variously long historical timeframes. In doing so, the paper contributes to ongoing debate over the conceptual and empirical status of failure in policy mobilities literature.

Introduction

On 31 October 2017, Daniel Andrews, the premier of the Australian state of Victoria, made an unexpected announcement: overturning his previous opposition, he announced the impending approval of a "Supervised Injecting Room" in North Richmond, an inner-urban neighbourhood of Melbourne. "This is a change in policy [...] no question about that", he stated, "[but] it's a change that is very much needed" (quoted in Barlow, 2017: n.p.). Upon the facility's opening, in July 2018, it became Australia's second Supervised Consumption Site (SCS)[1]. The change in policy coincided with a sharp increase in overdose deaths, centred in North Richmond, echoing the circumstances that surrounded the establishment of Australia's first SCS in 2001, the now internationally renowned Medically Supervised Injecting Centre (MSIC) in Sydney.

News coverage of the Melbourne SCS announcement emphasised that the recent advocacy efforts of several organisations and individuals had been decisive in the Premier's decision. A coalition including local business owners, medical professionals, police, community organisations, independent members of parliament, and the state coroner had created significant pressure in the years immediately preceeding the decision (Towell & Preiss, 2017: n.p.). Yet, the recent announcement cannot be disentangled from a much longer history of advocacy for SCSs in Melbourne. Greg Denham, executive director of the Yarra Drug and Health Forum, a key organization in the pro-SCS coalition, paid tribute to those whose efforts, over 20 years, paved the way for Melbourne's long-awaited SCS. "Despite many 'ups and downs'", he stated, "those committed to the cause [...] never lost confidence [...] that one day the 'planets would align' and such a centre would be set up" (Denham, 2017: n.p.). Given the history of that struggle, however, it seems clear that the effects of SCS advocacy – even when such advocacy appears to "fail"at certain points in time – are multiple, highly unpredictable, and often produce a wider set of harm reduction programs and services than just SCSs.

Since the 1990s, people who use drugs (PWUD), allied advocates like Denham, service providers, and policy-makers have developed and identified models that address the harms of illicit drug use in ways that counter dominant criminalization approaches and their accompanying stigmatization of PWUD. These harm reduction models include needle and syringe distribution, methadone prescription, and more advanced interventions like SCSs, which have been accepted as effective and replicable models by public health practitioners (EMCDDA 2017). Harm reduction is a philosophy and set of practical interventions developed to reduce the health, social, and economic harms of illicit drug use (e.g., from overdose and bloodborne disease, to drug-related litter, to the high costs of emergency room visits), without necessarily expecting reduced drug consumption or demanding abstinence.

SCS have been operating in European cities, including Bern, Frankfurt, and Amsterdam, since the mid-1980s and are also operating in Canada. They are either legally sanctioned or they operate in a legal grey zone. People bring drugs they have obtained elsewhere and consume them using equipment provided by trained staff. Drugs are consumed by injection, inhalation, or intranasally, depending on the regulations governing individual sites. They are observed by staff during and after consuming and they are often offered advice and resources. Despite the dangers of the unregulated drugs involved, there have been no deaths in the over 100 SCSs operating in 11 countries since 1986 (Global Platform for Drug Consumption Rooms, n.d.). The model is a travelling one, mobilised among cities in Europe and elsewhere by public health practitioners, PWUD, and their allies in response to shared problems associated with concentrated drug use in inner-urban and suburban locations – places drug policy researchers have described as "'drugscapes,' areas of cities [...] produced by social isolation and underdevelopment, where certain patterns of drug use are more likely to occur [and where] geographies of containment are created, enforced and reinforced" (Tempalski & McQuie, 2009, 7). Yet, SCSs differ from place to place depending on locally-specific drug use characteristics, regulatory frameworks, and public health problems (McCann, 2008, 2011b; McCann & Temenos, 2015).

Sydney's MSIC has been an important, if geographically isolated, reference point in global discussions of SCSs. In the lead-up to its establishment in 2001, the MSIC was expected to be one of three Australian SCSs, but similar initiatives in Canberra and Melbourne did not come to fruition. Thus, the Australian experience offers an example of the successful mobilization of the SCS model from its European heartland to the Southern Hemisphere, and a paradoxical example of what, until Premier Andrews' 2017 announcement, might be called a "failure" of attempts to mobilise and proliferate the SCS model across a wider range of cities. How do we understand the apparent inability of a globally-circulating model to be embedded and operationalised in a particular locality? In this paper, we address this question through the 20-year history of SCS proposals and their effects in Melbourne. In doing so, we make two contributions. First, we conceptualise policy "failure" in terms of its complex social, political, spatial, and, particularly, temporal contexts. Second, building on this perspective, we analyse "failure" in one historical moment to address how the politics of policy-making and the derailing of particular proposals may generate subsequent outcomes nonetheless.

As we argue in the next section, failure cannot be taken at face value or understood as a discrete condition. Highlighting the generative possibilities and pathways of apparent failure in policy-making and policy mobilization contributes to the ongoing debate over the conceptual and empirical status of "failure" in policy mobilities literature. To substantiate this argument, we provide an account of the politics of the Melbourne case in the subsequent section. This discussion explores the generative effects of SCSs proposals that did not come to immediate fruition. Rather than framing such thwarted attempts as categorical failures, we show how they not only inspired the recent apparently successful campaign for an SCS, but, over the past 15 years, they also altered the parameters of policy acceptability and invigorated policy and practitioner networks. The generative effects we discuss may be linear (a policy fails, actors learn from the failure, and a new modified policy is created) but we would suggest that, in most cases, policy-making and policy "failure" are defined by complex causalities and overdetermination. Therefore, the specific ways in which policy "failure" generates new strategies are contingent on the socio-spatial context in which policies are debated and reworked and, thus, they must be analysed and conceptualised in and through specific cases. In Melbourne, the post-"failure" outcomes led to the establishment of other services that many local actors believe were different, but appropriate, applications of harm reduction principles to their specific local situation. We conclude the paper by emphasising the political aspect of policy-making and policyfailing (Wells, 2014) – a theme that is evident throughout the paper but that is worth emphasizing in terms of its consequences for the analysis of policy mobilities.

Our discussion is based on analyses of documentary materials and key informant interviews. We reviewed documentary materials, including news media articles, government documents, and activist publications, from the last 20 years. Recurring themes were identified and the materials were used to provide an overview of defining debates and important actors. The documentary research supported a subsequent set of semi-structured interviews with 10 actors who were involved in policy debates and service delivery in the late 1990s, when SCSs were first proposed in Melbourne. Interviews were conducted in Melbourne in June 2016. Interviewees ranged from harm reduction advocates, social service workers, health researchers, a lawyer, a former police officer, and a senior politician, among others. Interviews were recorded and transcribed, then analyzed for recurrent themes in tandem with thematic analysis of documentary materials (Boyatzis, 1998; Rice & Ezzy, 1999). The analysis is further informed by one co-author's long-term international study of SCSs (Longhurst & McCann, 2016; McCann 2008, 2011; McCann & Temenos, 2015).

Failure and beyond

Accepting, to a greater or lesser extent, the premise that "[m]obility is an inherent characteristic of policy" (Freeman, 2012, 20), the policy mobilities literature seeks to characterise, conceptualise, and empirically investigate the ways in which policies and policy knowledge travel, how these travels create ever-incomplete similarities among diverse places, how policy-making is political, and how it is often as much a global as a local process (McCann, 2011a; McCann & Ward, 2011). In a recent discussion, Ward (2017) provides a valuable summary of the key elements of the policy mobilities approach. The literature foregrounds the ways in which places are produced through relations that run through and extend beyond them, as well as by the territorialised politics within them. While multiple types of relations constitute places, studying policy-making as relational place-making is a valuable contribution to both the study of place and policy. Drawing on the wider mobilities literature (Adey, Bissell, Hannam, Merriman, & Sheller, 2014; Sheller & Urry, 2006), the policy mobilities approach also takes seriously the interstitial spaces and activities among sites of policy development and adoption. Transit and translation, in this sense, are productive activities and, therefore, policy mobilities simultaneously involve mutations, both of the policies being circulated and of the places through which they circulate. This is a social process, so the literature has focused heavily on the various actors involved, from politicians, policy professionals, representatives of global organizations, and consultants, to social movement and community activists, among others. The focus on "transfer agents" in mobilizing policy emphasises "the embodied and performative nature of the work done [and] the importance of how people communicate and interact" as well as "the various objects, spaces, technologies and times that facilitate" mobilizations (Ward, 2017, 13). These actors work in, through, or in reference to institutions and informational infrastructures, particularly for the swift proliferation of hegemonic policy models (Peck & Theodore, 2015), but also for models that move at moderate speeds and, occasionally, for the spread of counterhegemonic ideas (Massey, 2011). Therefore, actors involved in promoting or questioning policy mobilization are continually engaged in some form of politics. Problematisation, comparison, "solutioneering", and persuasion, are all elements of the "supply side" politics of policy mobilities, taking place both at a distance and, crucially, in what Temenos (2016) calls the "convergence spaces" of conferences, fact-finding trips, etc. (see also Cook & Ward, 2012). As we will see, the impulse and imperative to mobilise policies is often met by resistance or skepticism, among other potential barriers.

The development of the policy mobilities literature over the last decade has been marked and facilitated by a rich "rolling conversation" (Peck, 2011) among various interlocutors. The literature has been challenged to better historicise the processes it studies (Harris & Moore, 2013), to focus both on the topologies, as well as topographies, of invention/adoption among locally-based urban actors who "arrive at" new policies (Robinson, 2013, 2015), and to broaden its scope beyond neoliberal policies and beyond the global North (Bunnell, 2015). Yet, perhaps the most persistent critique has been around the question of "failure". Is the literature "successist" (i.e., focused on success while blind to wider complexity)? Can it adequately account for policies that do not get mobilised, that come to nothing, that have unexpected or negative effects after they are implemented, or that continue to be mobilised after having fallen flat?

"Failure" as a focus of the policy mobilities literature

Given its constructivist orientation, "failure" and "success" are conceptualised as socially produced, spatially situated, and power-laden in the policy mobilities literature.[2] Failure and success are not natural, nor entirely distinct. They are relationally interconnected (McCann & Ward, 2015). For example, Ward (2006) notes that policies are "'made' into [...] success[es]," and, he continues, "there is nothing natural about which policies are constructed as succeeding and those that are regarded as having failed." Over its first decade, the literature has continued to acknowledge the role that constructions of failure play in the practice and politics of policy-making, while largely pursuing a very reasonable attention to policies that have moved and been implemented as the best way of fleshing out the conceptual contours of the approach (Baker & Temenos, 2015).

As a result of this dominant orientation, the appearance of failure in the literature has largely been in the form of critique. Critics have questioned what they see as an over-attention to successful examples of policy implementation and to the mobilization of successful policy models. Jacobs (2012), in one of the earliest critiques, calls for more attention to "[s]ites of failure, absence and mutation" (see also Clarke, 2012). This argument has emerged in parallel and, frequently, in combination with charges of problematic "presentism" (Harris & Moore, 2013; Jacobs & Lees, 2013; McFarlane, 2011), a term that can be read as connoting either an overt focus on the contemporary (neoliberalised) period (Bunnell, 2015) or, more pertinently to our discussion in this paper, a tendency to be attracted to cases where policies associated with one place are present elsewhere, suggesting their successful implementation. More recently, these critiques have gained momentum. "[I]t is necessary to pay attention to [...] interruptions, exceptions, and stalled attempts at policymaking," Wells (2014) argues. These are "moments in which policies are defeated, stopped, or stalled," she continues. Stein et al (2017, 36) go as far as to identify a "success bias" in the literature. They also "argue that it is necessary to examine failure, resistance and contradictions [by] focusing on breaks, cuts, stoppages and detours." Lovell (2017, 4), for her part, argues that "policy mobilities research is overwhelmingly about policies that do work and are 'present' – publicly promoted and discussed as successes."

Clearly, policy mobilities researchers should take failure seriously. After all, there are numerous cases where policies fail on their own terms. Stein et al. (2017) study the uneven fortunes, and frequent failures, of Business Improvement Districts in Germany, for example. They emphasise the various contextual conditions that can stymie the smooth "transfer" of a neoliberal policy fix. They rightly stress that policy mobilization is a "contested, fractured and often inherently contradictory process marked by unpredictable outcomes" (Stein et al., 2017, 28). Lovell (2017a) discusses a failed smart electricity metering program in the state of Victoria, Australia, and how, despite failing on its own terms, was discussed in other states, and then nationally, as a cautionary case to be explicitly disavowed. Some policies plainly do not achieve their stated or implied aims. Worse still, they can motivate consequences and reactions that further exacerbate the problem they are meant to address. Failure is an apt term to describe the reality of such occurrences.

Contextualizing, temporalizing, and differentiating failure

Policies are "embedded within [...] particular social and cultural worlds or 'domains of meaning' [that they] create as well as reflect" (Shore & Wright, 2011, 1–2). Therefore, studies of the limits of policy mobilization must extend beyond *prima facie* understandings of what failure is and analysis of these limits should avoid fetishizing, simplifying, dichotomizing, or reifying failure. Analyses can address how the effects of an attempt to implement a policy model are defined, either on their own terms or by their differential impacts on a range of interests and constituencies in specific social, spatial, and temporal contexts. As we will suggest in the following paragraphs, this nuanced approach is already evident in recent writing on failure in the literature. It is a promising orientation to which we hope to contribute through our case study, by considering the long-term after-effects of attempts to introduce the globally-mobile SCS model to Melbourne. A consideration of the complex effects of those thwarted attempts allows us to address what failure meant and did not mean in this specific case. The case also allows us to show how researching a relatively long historical timeframe influences our understanding of policy advocacy, policy mobilization, policy-making, and policy failure. In turn, it raises questions about which concepts and terms are most appropriate rubrics under which to address the fate of policy models in certain contexts.

These questions have been raised by others. For example, the brief and provisional critiques of the policy mobilities approach by Jacobs (2012) and McFarlane (2011) suggested that there was more to policy-making and policy mobilization than had been captured in the literature, which was very much in its infancy when they wrote. It is telling, for example, that Jacobs (2012), who critiques a tendency to focus on the repetition of policies rather than on their differentiation, argues that "[s]ites of failure" *but also of* "absence and mutation are significant empirical instances of differentiation." She then calls for analysis of "the motives and politics of action-in-the-name-of-differentiation, reaction, rejection, de-activation, detour, redirection and failure" (ibid.). The point, then, would be to pursue a "differentiation-focused" agenda in the study of policy mobilities.

Longhurst and McCann (2016), for example, examine the politics of harm reduction in Surrey, BC, Canada, a suburban municipality in Greater Vancouver. At the time of the study, the implementation of harm reduction models in the city was staunchly resisted by local politicians and police, despite being central to overarching Provincial health strategies. This resistance was set in particularly stark relief by Surrey's proximity to Vancouver, a city globally recognised as a successful model and mobiliser of harm reduction approaches. While Surrey's recalcitrance could be interpreted as evidence of a failure of the model's mobilization, the study suggests that definitions of "failure" are influenced by changing social and political contexts and by the ways in which research into the case is temporally delimited. Thus, the Surrey case supports a conceptualization of *constrained* policy mobility in which barriers, boundaries, and frontiers of mobilization are understood as spaces of alliance building, debate, persuasion, experimentation, and persistence as key actors wait for political and other circumstances to change. This argument has subsequently been borne out as a huge increase in overdose deaths and continued pressure from activists and public health officials led to the establishment of two SCSs in Surrey, despite the objections of local business and political leaders (Britten, 2016; Reid, 2017).

The timescale of one's analysis is, therefore, crucial in the study of policy mobilities and policy failure. Furthermore, it is problematic to entirely separate success and failure in these sorts of analyses (McCann & Ward, 2015). Wells' (2014) notion of "policyfailing," addresses both these points. Policymaking and policyfailing are two sides of the same coin, she argues: "systematic regulatory failing is endemic to governance and likely constitutes the actual essence of policy-making efforts" (ibid., 479) and, she continues, "policyfailing" is "an ongoing and unstable process rather than [...] an unequivocal achievement" (ibid., 488). Just as no policy model is forever, an apparent policy failure is not necessarily an end-state. The fate of a policy and the effects of failure are matters of power and politics, as Wells (2015) argues elsewhere. Competing agendas, not only between proponents and antagonists of a particular policy but also among these "camps", can both make a policy fail and also complicate, reformulate, and redirect its post-failure future. In Chang's (2017) study of an early attempt to build a model eco-city in Dongtan, "failure matters" because the context of her study is defined more widely (spatially and temporally) than simply a focus on the story of Dongtan, from its planning to its failure. Rather, by tracing the legacies and travels of Dongtan's failure, she is able to show its influence on a subsequent eco-city development elsewhere in China. Complicating the division between success and failure, while echoing Wells in her discussion of unpredictability and process, Chang argues "that a model may not be successful in its implementation but remains successful in its mobility" (ibid., 1721). Moreover, Chang's research shows that failure can generate and facilitate further learning, innovation in, and differentiation of policy. She notes that McFarlane (2011, 373-374, our emphasis) sees failed initiatives as helpful to learning because they can initiate "new habits of working and challenging regimes of truth, as well as building capacities of engagement [with] the potential of transformation, and of the emergence of a *different* kind of city."

As we have shown, the policy mobilities conversation has long been marked by attempts to develop a critical and nuanced approach to all aspects of policy-making and policy advocacy, including the question of failure. A number of scholars have pushed beyond thinking in terms of stark and unequivocal failure (set against studies that *supposedly* look for and find stark and unequivocal success). Instead, they explicitly refer to and grapple with questions of differentiation, mutation, fragility, unraveling, instability, emergence, detour, redirection, reaction, rejection, de-activation, and absence. This constellation of terms, drawn from the works discussed above, suggests that scholars are unpacking and differentiating failure in their analyses. It also highlights the centrality of time and temporality in politics and policy-making, as has been recently noted by anthropologists (e.g., Abrams, 2014; Harms, 2013) and geographers (e.g., Anderson, 2010; Bunnell, Gillen, & Ho, 2017; Wood, 2015) among others. In what follows, we approach an example of apparent failure in this way and add to the ongoing conversation by examining its long-term generative effects, where these effects are understood to be complex, non-linear, overdetermined and operating in co-constitutive relationships with their socio-spatial contexts.

Crisis politics: the fate of SCS proposals in Melbourne (1997–2002)

The most intense period of advocacy and public debate around proposals for SCSs in Australia occurred between 1997 and 2002, amid a sharp increase in rates of heroin overdose and overdose-related death in large Australian cities. In Canberra, Sydney and Melbourne, distinct but interconnected coalitions of community groups, politicians, health practitioners and researchers sought to respond to the unfolding public health crisis with a range of harm reduction initiatives (Dietze, Fry, Rumbold, & Gerostamoulos, 2001). These included the expansion of existing initiatives, such peer education and needle/syringe exchange programs, but also involved moves to implement new harm reduction initiatives, such as drug/alcohol recovery spaces, heroin prescription, and SCSs.

Proposals for SCSs were the focus of state-level political debates – in the Australian Capital Territory (ACT), New South Wales, and Victoria, respectively – because of the central role that state governments play in funding drug treatment services and creating the legal environment for SCSs to operate. In 2000, the proposal for a SCS in Canberra was wrecked on the shoals of budget negotiations between the sitting government and independent members of the ACT legislature (Gunaratnum, 2005). A public referendum was later considered but never held, and no significant attempts to implement a SCS have been made since. As we have noted above, the proposal for a SCS in Sydney was successful and the resulting Medically Supervised Injecting Centre (MSIC) opened in 2001 as a pilot program. It was subsequently given a longer mandate and at the time of writing (June 2018) continues to be the only operating SCS in Australia. Sydney's MSIC is widely regarded – domestically and internationally – as a highly effective public health intervention (KPMG, 2010). For a time, proposals for SCSs in Melbourne seemed to have sufficient momentum to lead to the creation of at least one. However, this was not to be, until the recent announcement of a plan to open one in North Richmond. "The policy process [in the late 1990s and early 2000s involved] a lot of actors and [was] politically complicated too," one senior health researcher recalled (Interview 8: health researcher). This complex process can be broken down, heuristically, into three phases.

Phase 1: the "heroin crisis"

The first, from 1997 to 1999, was characterised by the emergence of what was commonly called a "heroin crisis", which created political opportunities for advocates to suggest new public health initiatives and approaches. Melbourne's heroin-related deaths had increased from 49 in 1991 to 268 in 1998 and cases of non-fatal overdose increased significantly, resulting in "widespread recognition of the urgency of the problem of heroin overdose" (Dietze et al., 2001: 437). Increases in heroin-related harms were generally thought to have resulted from a "glut" of inexpensive, relatively pure heroin, high-risk drug use practices, and the emergence of open drug markets in several locations. The visibility and scale of heroin-related harm and the related "disorder" observed in public spaces encouraged, at least for a time, a view that the heroin crisis was one of public health, rather than criminality or individual fault (Fitzgerald, 2013). A senior figure within the Melbourne drug treatment sector described the situation:

> The mayhem of open drug markets was a new thing that we'd never had before and then the failure to actually maintain public amenity in relation to injecting equipment was important. There were needles strewn about [...] playgrounds and on beaches and public transport and shopping centres. [...] There was a level of sympathy in the community as well. The *Herald Sun* is a tabloid [Rupert] Murdoch paper and they used to run the overdose toll next to the road [accident death] toll [on the front page]. They were milking it in a way, but on the other hand, at least they were actually paying attention to it. (Interview 5: drug treatment professional)

The first SCS proposals came in 1997 as the heroin crisis worsened. In August 1997, the City of Greater Dandenong, in outer Melbourne, released a report advocating for the introduction of "Safety Clinics". One month later, Eddie Micallef, a member of the Victorian state legislature, proposed an SCS in the suburb of Springvale, within the same municipality. These proposals signalled the start of a protracted period of debate, research, policy development, and electoral positioning that incorporated state government and several municipalities within the metropolitan area. Individual councillors from the City of Melbourne and the City of Maribyrnong signalled their support for an SCS within their jurisdictions, while the City of Port Phillip and City of Yarra voted in favour of trial sites. One interviewee noted that "it took a year or two until the idea made its way into the state government political process – then it just gathered steam" (Interview 8, health researcher).

Separate from local and state government, a group of people associated with Wesley Central Mission, a church in the Central Business District, took it upon themselves to construct a SCS – at significant financial and reputational risk – by refurbishing a building on church grounds. Sensing that the political winds were shifting in support of one or more SCSs, their initial intention was to open the facility once local planning approval was granted and enabling legislation at the state level had passed (Interview 9: harm reduction advocate). When conservative Liberal Party premier Jeff Kennett's support for SCSs appeared to soften in the lead-up to the September 1999 state election, the Wesley Central Mission group met with the premier's staff and threatened to open the facility as an act of civil disobedience (a strategy used, over the years, by SCS advocates in Europe, Canada, and in Sydney (Wodak, Symonds, & Richmond, 2003)):

> We walk in and we meet [an advisor to Premier Kennett]. There's nine of us and one of him, God bless him, and we say we've made a decision: we're opening [the SCS] without regulation or legislation. We don't care what the consequences are. I've never seen a man's face change colour so fast. He leaves the room and is gone for forty minutes. He comes back in and he says, 'if [...] you don't do it [until] after the election, we will pilot one [if elected]'. (Interview 9: harm reduction advocate)

With an assurance from the premier, and widely-held expectations that the Liberal Party would be returned to government at the upcoming election, the Wesley Central Mission group agreed to wait and, as a result, a fully-functional SCS sat unused in the city centre.

Meanwhile, the opposition Labor Party sought to differentiate itself from Kennett's Liberal Party government and to capitalise on emerging professional and public consensus regarding the acceptability and necessity of SCSs. As part of their election platform, Labor announced a trial program with not one, but five SCSs across the Melbourne metropolitan area, pending community consultations in the areas surrounding the proposed site of each trial. In September 1999, opposition leader Steve Bracks led the Labor Party to an unexpected electoral victory, which saw the party form a minority government with the support of three independent members of the Legislative Assembly (the lower house of state parliament). The Legislative Council (the upper house) remained under the control of the Liberal and National parties.

Phase 2: the Labor Party's support for SCSs

The Labor government's subsequent attempt to implement trial SCSs, in the period from 1999 to 2000, marked the second phase in Melbourne's history of SCS proposals. With the SCS trial sites high on the new government's agenda, advocacy efforts quickly intensified on both sides of the issue, and the positions of politicians, community groups, and others hardened. Disagreement among Christian groups was particularly pronounced. Father Peter Norden, director of Jesuit Social Services, came out in support of SCSs, claiming that it was "very immoral for Christian people to stand by [during such a crisis.] Safe injecting rooms are not about encouraging heroin use, they are about keeping people alive" (cited in Duffy, 1999: 32). Melbourne's Catholic Archbishop George Pell took a different stance. Noting that involvement by Catholic organisations was "misdirected compassion", he reportedly used his influence in the Vatican to have a decree issued by the Congregation for the Doctrine of the Faith to ban Catholic groups from involving themselves in SCSs (Duffy, 1999: 32). Opinion was also divided in the local city councils, particularly the City of Melbourne, whose councillors were airing their disagreements in the media on whether to support or oppose the proposed SCS in the city centre (Strahan, 1999: 9).

With signs of an increasingly fractious public debate, and concerns about the new government's fragile parliamentary majority, Premier Bracks' "approach to governing became one of very small steps and a very conservative approach" (Interview 1: health researcher). He outsourced policy development and consultation to an independent Drug Policy Expert Committee, comprised of key actors within Victoria's drug research and drug treatment fields. The Committee was charged with conducting public consultations on the proposed SCS trial and providing wide-reaching recommendations for the implementation of SCSs and other harm reduction initiatives to address the heroin crisis. Instead of providing a release valve for the discontent of certain segments of the public, however, the consultations provided more opportunities to call the appropriateness of SCSs into question. Advocacy groups sprang up in response to specific trial sites. In one inner suburb, a supportive group, called Footscray Cares, and an oppositional group, called Footscray Matters, were heavily involved in efforts to sway local political and public opinion. While survey results released by the Drug Policy Expert Committee in April 2000 showed support from 64% of residents in the vicinity of the trial sites, rancorous public meetings and, at times, sensationalist media coverage amplified the impression of a deeply divided populace. As the Liberal opposition health minister at the time noted:

> in my mind, when Labor floated this idea they made a fatal political error: they announced where the sites were going to be. Politically, we'd pressed them on that [...] The community went ballistic. (Interview 7: politician)

Phase 3: from legislation to rejection

Nonetheless, in May 2000, the Labor government moved ahead by introducing a bill in the Victorian Parliament that, if passed, would have enabled five SCSs. This was the third phase of the process, which lasted until the five-site proposal was formally abandoned in 2002. While introduction of enabling legislation into parliament marked a significant milestone in the path toward implementation, it quickly became apparent that oppositional forces were growing and, in doing so, sapping much of the political momentum out of the proposed trial. In May 2000, the City of Melbourne rejected an application for planning approval by Wesley Central Mission for its SCS and, in June, the City of Greater Dandenong and Victoria Police both refused to support trial SCSs.

At the same time, the "heroin glut" had turned into a "heroin drought" in 2001. The sudden and prolonged shortage of heroin produced dramatic changes in Melbourne. Street-based drug markets were far less active and heroin-related harm reduced markedly. From December 2000 to January 2001, heroin-related deaths fell by 82% in the state of Victoria and non-fatal heroin overdoses declined by 52% in Melbourne (Dietze et al., 2004: xi). In the eyes of the public and key decision-makers, these changes merited the downgrading of the current situation from a previously recognised crisis to a more normalised state of affairs. Without widespread recognition of exceptional, critical circumstances, the justifications for radical proposals, such as SCSs, lost their potency:

> Once you didn't have what people regard as an unacceptable number of heroin related deaths, then they said 'well, you know, why would you have injecting rooms?' 'Why would you have places where people can go use their drugs, because that'll just encourage them'. It went back to some of the old arguments. (Interview 2: drug treatment service manager)

This point was echoed by another interviewee, who remarked that, "harm reduction strategies generally are more supported when there's a sense of emergency and the broader community is at risk" (Interview 6: advocacy organisation manager). The same interviewee noted that, during the heroin drought,

> the heroin trade became very much behind closed doors [and it] probably became more complex and more difficult to police and those street drug markets that had been in the city, and other places were generally pushed towards places where [they] could better survive.

As a result, by September 2000, the opposition Liberal Party decided to oppose the government's enabling legislation, and in November, the Legislative Council rejected the bill. The future of the proposed SCS trial remained in limbo until the lead-up to the state election in late 2002, when it was abandoned by a Labour government facing an impending "law and order" themed election campaign by the opposition Liberals.

Generative effects of SCS proposals in Melbourne

As we have already argued, failure is a real element of the politics of policy. As a model that ran counter to the hegemonic approach to people who used illicit drugs in Australia, SCS proposals had an extremely slim margin for error. Their fate is an example of the fragility Lovell (2017b) points to as a feature of policy assemblages. If we consider that there was little disagreement about the effectiveness of SCSs at the time, that they were proposed during a widely recognised public health crisis, and that there was, for certain time, bipartisan commitment to implementing one or more SCSs, the proposals for SCSs in Melbourne had much going in their favour. Labor's change of heart in 2002 marked a failure for proponents of the model and, in the immediate period, SCSs did not come to fruition.

Yet, a different set of insights emerge when we look back over the intervening 15 years, scoping out from the specific moment when the SCS proposals unravelled in Melbourne. With a longer historical view, it is apparent that simply labelling the fight for SCSs as a failed attempt at policy implementation would be both limited and limiting. It would assume, first of all, that there was a coherent, singular goal among harm reduction advocates in Melbourne at the time and that all these varied actors were true to that goal. While some actors were advocating solely for SCSs, others were arguing for a range of harm reduction programs that included SCSs alongside primary health facilities, heroin prescription, expanded drug treatment programs, and expanded needle/syringe exchange programs. This tends to be the case in harm reduction advocacy efforts: advocates usually support a continuum of care. As one interviewee noted, it is "a little bit more complicated than just saying, 'Oh, injecting rooms didn't get up [and running]'" (Interview 1: health researcher). Furthermore, many of the key advocates were pushing for SCSs in multiple Australian cities at the time and they framed their goals as nationally-scaled rather than Melbourne-specific. Therefore, for another interviewee, framing Melbourne-based advocacy as a failure makes little sense:

> It wasn't a Melbourne thing for us [...] We needed to get one in Australia [...] If you want to look at the victory of it, we [now] have the research that can contribute to the literature [because of the Sydney SCS]. I was very happy. (Interview 9: harm reduction advocate)

Secondly, designating the Melbourne case as categorical failure obscures a substantial amount of policy-making activity that continued after the fact. The unravelling of SCS proposals had discernible and lasting impacts. Focusing too narrowly on a single moment of setback presents a real risk of framing failure as the inert and passive "other" to generative and active success (McCann & Ward, 2015; Peck & Theodore, 2015; Wells, 2014). Instead, we can pursue the notion of "generative failure". Peck and Theodore (2015: 140) note, for example, that "neoliberal interventions tend to 'fail forward'". Indeed, their work suggests that failure-induced policy experimentation is a central feature, rather than an exception or fatal flaw, of contemporary policy-making. While neoliberalization is not the focus of our discussion, per se, we suggest that an attention to the generative effects of setbacks and rejections is a useful way of nuancing the ongoing "failure debates" around policy mobilities.

We sought out the Melbourne case because it provides an opportunity to critically interrogate the implication, in many critiques of the policy mobilities approach, that policy failure has no impact and, therefore, no importance to policy studies, except as a cautionary tale or a case from which to draw negative lessons. Our intention was to return to the scene, some 15 years after Bracks' Labor government abandoned the implementation of SCSs, to assess whether there were any generative effects associated with the unsuccessful SCS proposals. We found two main ways in which the proposals were generative.

Detour, redirection, and the generation of other programs

The first way that failed SCS proposals were generative was that serious discussion of them, research into them, and even the physical "modeling" of one of them in the city centre expanded the parameters of policy acceptability for other harm reduction initiatives. This expanded universe of potentially acceptable options led, most immediately, to the establishment of five primary healthcare facilities for people who use drugs, the first of their kind in the state of Victoria. Located in the five locations that had been ear-marked for SCSs, the facilities – which provide services such as needle and syringe distribution, peer support, counselling and drug treatment – were funded by the budget allocation that had been reserved for the SCSs. These programs were a concession to proponents of SCSs, and their contribution to the health of people who use drugs and others in Melbourne has been significant. Indeed, several interviewees believed that, in retrospect, these centres were a better long-term response to drug-related harm than SCSs might have been. The dissipation of mature street-based drug markets and drug use in the wake of the heroin drought and the emergence of street scenes in different locations, meant that many of the SCSs would have, in the words of two interviewees, seemed like "white elephants" (Interviews 1 and 10) because of their under-use. Another interviewee explained,

> You might see [the primary healthcare centres] as the consolation prize [...] I'm more of the view that it was a better outcome. [...] the market changed so much in Melbourne in 2001. Barely after the [SCS trial] Bill was voted down, the heroin overdose rate dropped [...] People talk about the heroin drought. It was a shifting market. [...] The public injecting scene practically disappeared overnight. (Interview 8: health researcher)

Moreover, the executive officer of a peer-run harm reduction organisation for people who use drugs explained that, in the centres,

> the clinical staff [are] supported by peer workers: trained people who were also injecting drug users [...] it really was quite specific and unique to these primary healthcare facilities. It doesn't exist anywhere else [in Victoria] to this day (Interview 10: harm reduction advocate).

The interviewee explained that this approach allowed for a movement away from a "purely clinical model of healthcare" to a less alienating and more inclusive model that draws people who use drugs closer to the services they need to survive.

SCS proposals created an environment in which less radical proposals seemed palatable, in other words. In this way, advocates of SCSs employed a strategy – knowingly or not – that is commonly associated with libertarian and conservative think tanks (Peck, 2006). They advanced a radical demand, in the form of SCSs, that cast less controversial harm reduction initiatives as reasonable and feasible, thereby shifting the parameters of policy acceptability. While the primary health centres represent a direct and tangible outcome of the failure to implement SCSs, it is also reasonable to surmise that SCS advocacy – in creating what Peck (2006: 681) refers to as a "shift in the ideational climate" – had more diffuse impacts on the perceived acceptability of harm reduction initiatives. High profile public debates about the implementation of SCSs were more than debates about SCSs. By their very nature, they were about how drug use ought to be thought of and acted upon. They called upon politicians and the general public to question received understandings of drug use and people who use drugs. It is very likely that several years of public advocacy for SCSs contributed to the general acceptance, on the part of significant segments of the public, of much of the harm reduction "policy toolkit".

Capacity-building, engagement, and the generation of ongoing harm reduction networks

The second way that failed SCS proposals were generative was by acting as a focal point for the invigoration of harm reduction policy activism and policy networks. The urgency of the late 1990s heroin crisis compelled a range of people from various professional and personal backgrounds to become public advocates for harm reduction. Particularly noticeable was the extent to which SCS proposals galvanised community-based organisations, some of which continue to play prominent roles in harm reduction advocacy. The Yarra Drug and Health Forum, for example, rose to prominence through its advocacy for SCSs in the late 1990s. "There was some effective mobilization of community action groups as key participants in the drug policy debate," one interviewee noted. "There were," he continued,

> the opposition groups, but there were also some coalitions and supportive groups. ...
> [T]he debate that happened did breathe a bit of life into [the Yarra Drug and Health
> Forum.] We saw a higher order of participation in drug policy in Melbourne and
> Victoria in the action groups participating, including users. (Interview 8: health
> researcher)

After a brief period of inaction following the abandonment of the five SCS trial sites in 2002, the Yarra Drug and Health Forum has been central to all subsequent calls for SCSs in Melbourne. The most recent attempts in the neighbourhood of Richmond, led by the Forum, are informed by the institutional knowledge of prior attempts.

Beyond the catalysing effect that SCS proposals had on community-based activism, the proposals also prompted the creation and strengthening of connections with harm reduction researchers and activists across Melbourne, but also nationally and internationally. News coverage and the accounts of interviewees suggest that the late 1990s and early 2000s was a period of intense policy activity. Fact-finding trips to Switzerland and other countries in Europe were common among many of the key people within the SCS debate, and experts from those places were brought to Melbourne as part of the lobbying effort. Recounting the visits by Robert Haemmig and Franz Trautmann, two key figures in early SCSs in Europe, an interviewee talked about the lasting relationships – professional and personal – he had maintained with colleagues in the international network of harm reduction advocates (Interview 9: harm reduction advocate). Many people who were involved in Melbourne SCSs debates remain involved in research, activism and "formal" politics, and they are enmeshed in drug policy networks nationally and internationally. The future trajectories of numerous people were forged in the context of a particularly heady period for drug use in Melbourne. In concert with a range of other factors, SCS proposals have moved people toward the pursuit of harm reduction goals and inserted them into evolving networks of harm reduction expertise and activism.

Conclusion

Policy mobilities scholars have begun to grapple with the conceptual and empirical status of failure as a complement to analyses of successful and successfully-mobile policies. Focusing on the 20-year history of apparently failed SCS proposals in Melbourne, we have contributed to this collective effort by attending to the long-term, generative effects of a thwarted attempt to replicate the globally-circulating Supervised Consumption Site model in this particular place. While acknowledging that failure is real, in the sense that policies and policy mobilisations can fail on their own terms, we take the position that "failure" is also constructed within evolving and relational contexts. Failed attempts at policy mobilisation can spawn allied proposals, offer valuable lessons and experiences in the careers of different policy actors, and forge consequential connections within professional and activist networks from the local to the global. Thus, as Perrons and Posocco (2009, 132) argue,

> failure opens up key trajectories of globalising processes to critical scrutiny and investigation, as the "global" is understood, interpreted and experienced through stoppages in flows, cuts in networks and new forms of exclusion. Further, this focus on failure also suggests the importance of a close consideration of the configurations of the social which underpin the failure of globalising processes, the historical trajectories leading to them and the range of important re-alignments and specifications [that] enrich more abstract and generic accounts of globalisation.

Certainly, the effort to establish SCSs in Melbourne, and in the other Australian cities, tied harm reduction advocates there into global circuits of knowledge (McCann & Temenos, 2015; Van Beek, 2004). But their failure was also generative. It influenced the individual careers of harm reduction advocates and experts, and supported the further development of local, national, and global networks of policy and activist knowledge and a longstanding commitment to the cause (Denham, 2017: n.p.) of harm reduction that has influenced the character and practice of drug policy, drug treatment, and drug activism more generally. Therefore, the disappointments harm reduction advocates experienced in the early 2000s are balanced by an optimism derived from their subsequent resilience, flexibility, and pragmatism. When looked at as a form of "slow policy-making", failure looks different from 15 years in the future.

The recent opening of North Richmond's Medically Supervised Injecting Room in 2018 could be seen as a poetic conclusion to 20 years of SCS advocacy in Melbourne. Yet, the story we have told suggests that it is simply another stage in a continuing political struggle to fully apply a public health framework to the wellbeing of people who use drugs. It emphasises the political nature of policy mobilization, policy-making, policyfailing, and policy change. These processes involve ideological struggles between different interests and coalitions as they seek to define the appropriate future for a place and its citizens. They play out in often complex and sometimes contradictory ways (Wells, 2015) and are part of the differentiation of policies, places, and social relations (Jacobs, 2012). The frontiers of a policy model's circulation may be defined in terms of failure, therefore, but rather than seeing failure as decontextualised, discrete, and functionally inert, it must be understood in social, political, spatial, and temporal context. If "frontiers have a complex geography whose very outlines are the products of contestation," as Leitner, Peck, & Sheppard (2007, 311–312) argue, then the case of SCSs and harm reduction in Melbourne suggests that care must be taken in how we define our terms and the scope of our study if we are to adequately assess the character and consequences of failure in specific cases of policy-making.

Notes

1. We use the "Supervised Consumption Site (SCS)" in this paper as a generic term that refers to a range of facilities "where illicit drugs can be used under the supervision of trained staff" (EMCDDA (European Monitoring Centre for Drugs and Drug Addiction), 2017, 1). The specific character of these sites depends on the contexts in which they operate. In Australia, among other places, the focus is on injection. Therefore, terms like "Medically Supervised Injecting Centre" or "Supervised Injection Facility" are used. The argument we make in the paper would apply to a range of consumption facilities, however, so we use this generic term. Other generic terms exist: "Drug Consumption Room" (DCR) is common because it encompasses locations where inhalation and intranasal consumption are permitted (Global Platform for Drug Consumption Rooms, n.d.). Yet, it does not emphasise the importance of supervision. "Supervised Consumption Facility" is a common substitute as a result. But "facility" tends to suggest a particularly formalised institutional setting, which is not the case everywhere. The alternative, more geographical, "site," which has recently become common in Canada (Health Canada, 2017), overcomes the narrow connotation of "facility" and emphasises how the specific character and location of a SCS in a particular neighbourhood has a great deal to do with its effectiveness.
2. Our use of scare quotes around "failure" and "success" thus far in our discussion indicates the indeterminacy of the terms and the need to critically unpack them. Having made the point, we will largely dispense with the quotation marks.

Acknowledgments

We are grateful to all those who agreed to be interviewed. Thanks also to Simone Cooper for research assistance and to Cristina Temenos and John Lauermann for comments on an earlier draft. Helpful comments from the anonymous reviewers and editorial advice from Susan Moore are also greatly appreciated. This research was funded by grants from the Social Sciences & Humanities Research Council of Canada (435–2013–2197) and the School of Environment at the University of Auckland. We dedicate this article to the memory of Jenny Kelsall, the late Executive Officer of Harm Reduction Victoria.

Disclosure statement

No potential conflict of interest was reported by the authors.

Funding

This work was supported by the Faculty of Science, University of Auckland [N/A];Social Sciences and Humanities Research Council of Canada [435–2013–2197];

References

Abram, Simone. (2014). The time it takes: Temporalities of planning. *Journal of the Royal Anthropological Institute, 20*(S1), 129–147.

Adey, Peter, Bissell, David, Hannam, Kevin, Merriman, Peter, & Sheller, Mimi (Eds.). (2014). *The Routledge handbook of mobilities.* New York: Routledge.

Anderson, Ben. (2010). Preemption, precaution, preparedness: Anticipatory action and future geographies. *Progress in Human Geography, 34*(6), 777–798.

Baker, Tom, & Temenos, Cristina. (2015). Urban policy mobilities research: Introduction to a debate. *International Journal of Urban and Regional Research, 39*(4), 824–827.

Barlow, Karen (2017). Melbourne to get 'very much needed' Safe injecting room trial. *Huffington Post* (Australia). Retrieved 2018, 24 January from http://www.huffingtonpost.com.au/2017/10/30/melbourne-to-get-very-much-needed-safe-injecting-room-trial_a_23261299/

Boyatzis, Richard E. (1998). *Transforming qualitative information: Thematic analysis and code development.* Thousand Oaks, CA: Sage.

Britten, Liam (2016). Fraser health wants supervised injection site in Surrey, but mayor hesitant. *CBC News British Columbia.* Retrieved 2018, January 23 from www.cbc.ca/news/canada/british-columbia/supervised-injection-site-1.3684789

Bunnell, Tim. (2015). Antecedent cities and inter-referencing effects: Learning from and extending beyond critiques of neoliberalisation. *Urban Studies, 52*(11), 1983–2000.

Bunnell, Tim, Gillen, Jamie, & Ho, Elaine. (2018). The prospect of elsewhere: Engaging the future through aspirations in Asia. *Annals of the American Association of Geographers*, 108, (1): 35-51.

Canada, Health. 2018. Applying for a supervised consumption site. Retrieved 2017, 23, January from https://www.canada.ca/en/health-canada/services/substance-abuse/supervised-consumption-sites.html

Clarke, Nick. (2012). Actually existing comparative urbanism: Imitation and cosmopolitanism in North-South Partnerships. *Urban Geography, 33*(6), 796–815.

Cook, Ian R, & Ward, Kevin. (2012). Conferences, informational infrastructures and mobile policies: The process of getting Sweden 'BID ready'. *European Urban and Regional Studies, 19*(2), 137–152.

Denham, Greg (2017). Open letter of thanks: SIF announcement. *Yarra drug and health forum.* Retrieved 2017, 15 January from http://www.ydhf.org.au/sif_thanks_letter.html

Dietze, Paul, Fry, Craig, Rumbold, Greg, & Gerostamoulos, Jim. (2001). The context, management and prevention of heroin overdose in Victoria, Australia: The promise of a diverse approach. *Addiction Research and Theory, 9*(5), 437–458.

Dietze, Paul, Miller, Peter, Clemens, Susan, Matthews, Sharon, Gilmour, Stuart, & Collins, Linette. (2004). *The course and consequences of the heroin shortage in Victoria.* Australia: Turning Point and the National Drug and Alcohol Research Centre, University of New South Wales.

Duffy, Michael (1999). Morals and charity in collision. *Courier Mail.* 30 October.

EMCDDA (European Monitoring Centre for Drugs and Drug Addiction) (2017)Drug consumption rooms: An overview of provision and evidence. Retrieved2018, January23from http://www.emcdda.europa.eu/topics/pods/drug-consumption-rooms

Fitzgerald, John. (2013). Supervised injecting facilities: A case study of contrasting narratives in a contested health policy arena. *Critical Public Health, 23*(1), 77–94.

Freeman, Richard. (2012). Reverb: Policy making in wave form. *Environment and Planning A, 44*(1), 13–20.

Global Platform for Drug Consumption Rooms (n.d.). Retrieved 2018, January 23 from http://www.salledeconsommation.fr/index.html

Gunaratnum, Praveena. (2005). *Drug policy in Australia: The supervised injecting facilities debate.* Australia: Asia Pacific School of Economics and Government, Australian National University.

Harms, Erik. (2013). Eviction time in the New Saigon: Temporalities of displacement in the rubble of development. *Cultural Anthropology, 28*(2), 344–368.

Harris, Andrew, & Moore, Susan. (2013). Planning histories and practices of circulating urban knowledge. *International Journal of Urban and Regional Research, 37*(5), 1499–1509.

I-Chun Catherine, Chang. (2017). Failure matters: Reassembling eco-urbanism in a globalizing China. *Environment and Planning A, 49*(8), 1719–1742.

Jacobs, Jane M. (2012). Urban geographies I: Still thinking cities relationally. *Progress in Human Geography, 36*(3), 412–422.

Jacobs, Jane M, & Lees, Loretta. (2013). Defensible space on the move: Revisiting the urban geography of Alice Coleman. *International Journal of Urban and Regional Research, 37*(5), 1559–1583.

KPMG. (2010). Further evaluation of the medically supervised injecting centre during its extended trial period (2007–2011). Retrieved 2018, January 23 from http://www.health.nsw.gov.au/mentalhealth/programs/da/Documents/msic-kpmg.pdf

Leitner, Helga, Peck, Jennifer, & Sheppard, Eric. (2007). Squaring up neoliberalism. In (Eds.), *Contesting neoliberalism: urban frontiers*. New York: Guilford Press.

Longhurst, Andrew, & McCann, Eugene. (2016). Political struggles on a frontier of harm reduction drug policy: Geographies of constrained policy mobility. *Space and Polity*, *20*(1), 109–123.

Lovell, Heather. (2017a). Are policy failures mobile? An investigation of the advanced metering infrastructure program in the state of Victoria, Australia. *Environment and Planning A: Economy and Space*, *49*(2), 314–331.

Lovell, Heather. (2017b). Policy failure mobilities. *Progress in Human Geography*, 0309132517734074.

Massey, Doreen. (2011). A counterhegemonic relationality of place. In Eugene McCann & Kevin Ward (Eds.), *Mobile Urbanism: Cities and policymaking in the global age* (pp. 1–15). Minneapolis: University of Minnesota Press.

McCann, Eugene. (2008). Expertise, truth, and urban policy mobilities: Global circuits of knowledge in the development of Vancouver, Canada's "four pillar" drug strategy. *Environment and Planning A*, *40*, 885–904.

McCann, Eugene. (2011). Points of reference: Knowledge of elsewhere in the politics of urban drug policy. In Eugene McCann & Kevin Ward (Eds.), *Mobile urbanism: Cities and policymaking in the global age* (pp. 97–122). Minneapolis: University of Minnesota Press.

McCann, Eugene. (2011a). Urban policy mobilities and global circuits of knowledge: Toward a research agenda. *Annals of the Association of American Geographers*, *101*(1), 107–130.

McCann, Eugene, & Temenos, Cristina. (2015). Mobilizing drug consumption rooms: Inter-place networks and harm reduction drug policy. *Health & Place*, *31*, 216–223.

McCann, Eugene, & Ward, Kevin (Eds.). (2011). *Mobile Urbanism: Cities and policymaking in the global age*. Minneapolis: University of Minnesota Press.

McCann, Eugene, & Ward, Kevin. (2015). Thinking through dualisms in urban policy mobilities. *International Journal of Urban and Regional Research*, *39*(4), 828–830.

McFarlane, Colin. (2011). *Learning the city: Knowledge and translocal assemblage*. Oxford: Wiley-Blackwell.

Peck, Jamie. (2006). Liberating the city: Between New York and New Orleans. *Urban Geography*, *27*(8), 681–713.

Peck, Jamie. (2011). Geographies of policy: From transfer-diffusion to mobility-mutation. *Progress in Human Geography*, *35*(6), 773–797.

Peck, Jamie, & Theodore, Nik. (2015). *Fast policy: Experimental statecraft at the thresholds of neoliberalism*. Minneapolis: University of Minnesota Press.

Perrons, Diane, & Posocco, Silvia. (2009). Globalising failures. *Geoforum; Journal of Physical, Human, and Regional Geosciences*, *40*(2), 131–135.

Reid, Amy (2017). Health Canada approves two safe consumption sites in Surrey. *BC local news*. Retrieved 2018, January 23.from https://www.bclocalnews.com/news/health-canada-approves-two-safe-consumption-sites-in-surrey-2/

Rice, Pranee L, & Ezzy, Douglas. (1999). *Qualitative research methods: A health focus*. Melbourne: Oxford University Press.

Robinson, Jennifer. (2013). "Arriving at" urban policies/the urban: Traces of elsewhere in making city futures. In Ola Söderström, S. Randeria, D. Ruedin, G. D'Amato, & F. Panese (Eds.), *Critical mobilities*. Oxford: Routledge.

Robinson, Jennifer. (2015). "Arriving at" urban policies: The topological spaces of urban policy mobility. *International Journal of Urban and Regional Research*, *39*(4), 831–834.

Sheller, Mimi, & Urry, John. (2006). The new mobilities paradigm. *Environment and Planning A*, *38*(2), 207–226.

Shore, Cris, & Wright, Susan. (2011). Conceptualising policy: Technologies of governance and the politics of visibility. In Cris Shore & Susan Wright (Eds.), *Policy worlds: Anthropology and the analysis of contemporary power*. New York: Berghahn Books.

Stein, Christian, Michel, Boris, Glasze, Georg, & Pütz, Robert. (2017). Learning from failed policy mobilities: Contradictions, resistances and unintended outcomes in the transfer of "Business Improvement Districts" to Germany. *European Urban and Regional Studies, 24*(1), 35–49.

Strahan, Nicole (1999). Mayor backs safe drug room trials. *The Australian*. 27 October.

Temenos, Cristina. (2016). Mobilizing drug policy activism: Conferences, convergence spaces and ephemeral fixtures in social movement mobilization. *Space and Polity, 20*(1), 124–141.

Tempalski, Barbara, & McQuie, Hilary. (2009). Drugscapes and the role of place and space in injection drug use-related HIV risk environments. *International Journal of Drug Policy, 20*(1), 4–13.

Towell, Noel, & Preiss, Benjamin. (2017, October). Melbourne heroin injecting room trial gets green light. *The Age, 31*, 2017.

Van Beek, Ingrid. (2004). *the eye of the needle: Diary of a medically supervised injecting centre*. Crows Nest, NSW: Allen & Unwin.

Ward, Kevin. (2006). "Policies in Motion", Urban management and state restructuring: The trans-local expansion of business improvement districts. *International Journal of Urban and Regional Research, 30*(1), 54–75.

Ward, Kevin. (2018). Policy mobilities, politics and place: The making of financial urban futures. *European Urban and Regional Studies, 28* (3): 266–283.

Wells, K. J. (2015). A housing crisis, a failed law, and a property conflict: The US urban speculation tax. *Antipode, 47*(4), 1043–1061.

Wells, Kevin. (2014). Policyfailing: The case of public property disposal in Washington, D.C. *ACME: an International E-Journal for Critical Geographies, 13*(3), 473–494.

Wodak, Alex, Symonds, Ann, & Richmond, Ray. (2003). The role of civil disobedience in drug policy reform: How an illegal safer injection room led to a sanctioned, 'medically supervised injection center'. *Journal of Drug Issues, 33*(3), 609–623.

Wood, Astrid. (2015). Multiple temporalities of policy circulation: Gradual, repetitive and delayed processes of BRT adoption in South African cities. *International Journal of Urban and Regional Research, 39*(3), 568–580.

Playing with time in Moore Street, Dublin: Urban redevelopment, temporal politics and the governance of space-time

Niamh Moore-Cherry (iD) and Christine Bonnin (iD)

ABSTRACT

Whether urban redevelopment is considered a "success" or "failure" is dependent on the temporal framings that we privilege. Until relatively recently, geographers have neglected the temporal politics that underpin urban redevelopment despite space-time being a crucial aspect framing the urban experience under capitalism. In this paper we argue for a focus on temporal politics or the politics associated with how time is experienced. Drawing on a case study of a market streetscape from Dublin (Ireland), we argue that cities need to be understood as shaped by multiple, fluid and contingent temporal framings and temporalities. Secondly, despite attempts by various stakeholders to control time and timeframes for particular ends, our case study highlights the impotence of planning and the challenges both time and urban temporalities raise for urban governance. Both space and its temporal framings are fraught with contestation, complicating any potential analysis of urban policy success and failure.

Introduction

Traditional urban markets are public assets contributing substantial social value to cities, yet they are globally coming under threat from urban redevelopment agendas (Dines, 2007; Watson & Studdert, 2006; Watson & Wells, 2005). While urban regeneration projects and schemes promote organic, food and farmers markets as new arenas of gentrified consumption, the continued existence of customary marketplaces within the urban fabric is precarious at best and time-limited at worst (Gonzalez & Waley, 2013). Their presence is a throwback to traditional, historic forms of urbanism, at a time when re-branding aims to generate different urban futures. Yet, the significance of these retail nodes to everyday heritages and urban livelihoods – how people strive to make a living in the city – has made this "re-branding" transition less than straightforward (Gonzalez & Dawson, 2015). How urban pasts, presents and futures intersect and are managed has been brought into sharp focus. This paper explores the Moore Street market and its general environs in Ireland's capital city, Dublin. While this iconic street market has resiliently persisted in the urban landscape through the diurnal activity of traditional trading, its place within the contemporary city has been contested for over 30 years. This coincides with a period of intense property-led redevelopment in the city and an aggressive entrepreneurialism in urban governance (MacLaran & Kelly, 2014), evident in large-scale regeneration schemes such as the Dublin docklands, Smithfield and parts of the Liberties (Moore-Cherry, Crossa, & O'Donnell, 2015).

The Moore Street market has long been a feature of the streetscape and the lived heritage of Dublin's north inner-city environment. Moore Street is situated adjacent to the main thoroughfare, O'Connell Street (Figure 1), which has undergone extensive regeneration since the millennium. Functionally a transitional location, Moore Street marks the boundary between areas of intense redevelopment since the early 1990s and disinvestment, some of it due to speculative land hoarding as a major developer amassed a significant landbank, discussed later. The market is the city's oldest and longest lasting open-air street market, dating to the mid-eighteenth century. Over the last four decades, a series of redevelopment proposals for the area have bound the future of the market and the street closely together but none were implemented. In recent years, the situation has been made more complex by the emergence of an increasingly politicised heritage struggle on the street, privileging a narrowly-defined heritage and temporality. From buildings 14–17 Moore Street, Nationalist leaders of the Easter Rising of 1916 – an armed rebellion against English rule in Ireland, and some would argue the first step on Ireland's path to independence – surrendered to British Crown forces. It is this particular heritage that has become critical in recent debates around the future of the street. Simultaneously, as the urban effects of the Great Recession of 2008 took hold in Ireland (O'Callaghan, Kelly, Boyle, & Kitchin, 2015), a deepening neoliberalism of urban governance gained momentum. Yet in Moore Street, this process was thwarted for a variety of reasons, underpinned – we argue – by the diverse temporalities co-existing in this particular space.

While a politicised heritage struggle provides the context for the paper and is contributing to the indeterminacy of Moore Street, we do not intend to present a critical perspective on "heritage" and its discourses. Rather we utilize the particular case

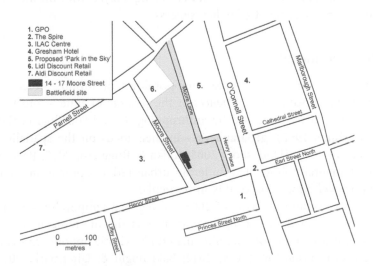

Figure 1. Location map of moore street and "battlefield" site.

study to rethink the place of time and temporality more generally in the politics of urban redevelopment. We argue that the temporal politics of the city is more than just the politics of time or unfolding of political events over time – *it is the formal and informal politics and contestations that emerge associated with how time is experienced*. In the paper we *firstly suggest that* cities need to be understood not through the linear, fixed, progress-oriented timeframes of bureaucratic stakeholders (Raco et al., 2008) but as shaped by multiple, fluid and contingent temporal framings and temporalities. The city is an arena where the past, present and future are variably brought into conversation with each other with differential material and socio-cultural consequences. *Secondly*, we argue that despite attempts by various urban stakeholders to control time for particular ends (Wallace, 2015), our case study highlights the impotence of planning and the challenges of governing brought about through variable interpretations and understandings of urban temporalities (Bastian, 2014). Whether planning or particular policies are perceived to be a "success" or "failure" can be highly dependent on the temporal framings that are privileged.

A qualitative approach was adopted for the study, comprising content and thematic analyses of key policy documents, plans, transcripts of protest speeches by national politicians and campaigners, and media reports. Semi-structured interviews were undertaken with key stakeholders including a local authority planner involved with regeneration plans for the area, three local politicians, one national politician, and representatives of two heritage campaign groups. Attempts were made to interview central government officials but they declined due to the sensitivity of legal proceedings around the site. Conversational interviews were undertaken with three traditional street traders whose families have been operating on the street for many decades, as well as with five newer migrant traders – all Chinese nationals – who run small shops along the street. We also undertook conversational interviews with three long-term local residents. The paper begins with an interrogation of the temporal politics of urban redevelopment. This is followed by a short discussion of the planning and heritage context of Moore Street, before the substantive discussion or "thick description" of the way different temporalities have been, and are being, played with on the street and the implications for the assessment of policy success.

Temporality and the city

In this paper, we focus on the urban temporalities of Moore Street – subjective, multiple, fluid and contingent *experiences* of the passage of time – and their inherent politics. Although widely discussed by anthropologists and sociologists (Degen, 2017; Harms, 2013; Munn, 1992), geographers – in their focus on the spatiality of place – have, until very recently, (Dodgshon, 2008; Kaika & Ruggiero, 2015; Raco et al., 2008) neglected the temporal politics that underpin urban redevelopment and place-making, and the *experiential* dimension of this in particular.

Since the global economic crisis of 2008, timing and temporality within the urban environment have gained increasing prominence as the unpredictability and cyclical nature of capital accumulation became materially visible in our cityscapes. In the context of austerity urbanism (Peck, 2012; Sakizlioglu & Uitermark, 2014; Wallace, 2015) in some European and North American cities, a sense of uncertainty about the future has come to permeate planning and policymaking. Attempts have been made to find temporal fixes to displace this most recent crisis of capitalism including a trend towards short-term solutions and temporary use activities (Bishop & Williams, 2012; Moore-Cherry, 2017; Till and McArdle, 2016). This has occurred in a bid to sustain the neoliberal city through difficult times (O'Callaghan & Lawton, 2016) and to minimise the visibility of policy failures.

This shift in the conceptualization of urban "development" highlights the centrality of time in both sustaining capital accumulation (Jessop, 2006; Kaika & Ruggiero, 2016) and in working through its crises. Space-time is a crucial aspect "framing the urban process and the urban experience under capitalism" (Harvey, 1985, p. 1). Yet, the politics of space-time and struggles to control time have received comparatively little attention (Raco et al., 2008). The power to initiate, halt and re-initiate projects is critical to the urban governance of redevelopment (Henderson, Bowlby, & Raco, 2007; Raco et al., 2008; Sakizlioglu & Uitermark, 2014; Wallace, 2015) reinforcing existing class-based power relations within the city. This control of time is inherently political as powerful elites generally dictate "at what point(s) in the development process different groups and interests could and should have their needs and priorities addressed" (Raco et al., 2008, p. 2652). The stalling of projects and emergence of "limbo-lands" (Wallace, 2015), one material expression of the rupturing of time – and the failures of neoliberal models of urban development – by the recent Great Recession highlights the importance of timing in continued neoliberal urbanization. In much the same way, contingency or temporariness has become a key asset in contemporary urban politics (Lauermann, 2016; Moore-Cherry & McCarthy, 2016).

While the *production and enactment* of time and timeframes and the sequencing of action and inaction is important, here we examine how powerfully productive and complex the *experiences of time* or temporal politics can be within contemporary urban redevelopment. Unlike the city of planning and planners, marked by "punctuated time" or event-driven time-frames (Guyer, 2007) such as redevelopment milestones, construction schemes and target setting, the experienced city is one of multiple temporalities. Others have argued that dominant spatial imaginations of particular urban spaces are produced through the control and delineation of "timeframes" and time in space (Kaika & Ruggiero, 2015; Raco et al., 2008). We argue that the urban environment is not just a technical-rational system following a linear path from past to present or present to future. Rather, it is an assemblage of multi-dimensional and superimposed cyclical, alternating and linear rhythms and diverse temporalities (Castells, 1997; Santos, 2006) that are the product of everyday activity and routinized action performed within particular socio-cultural and historico-political contexts and structures (c.f. Lefebvre, 2014). In order to disrupt and resist traditional power relations, Santos (2006) argues for "a new time literacy" (2006, p. 23) that embraces non-linearity, polichronicity and continuity. In our case study of Moore Street and its market, we highlight how such multiple temporalities prevent the assertion and follow through of a dominant spatial imagination. While these temporalities are critical in resisting the implementation of an entrepreneurial development agenda, they also produce new elites that complicate the framing and performance of urban governance.

This diversity of temporal framings deployed by a variety of stakeholders, linking present to pasts, and different pasts to different futures, are thus productive of formal and informal politics. The experienced city is defined *through* space and time (Crang, 2001) creating friction. As the site "where multiple temporalities collide", our case study of proposed redevelopments of Moore Street and its market illustrates the politics produced when "a particular constellation of temporalities [come] together in a concrete place" (Crang, 2001, p. 189–90). Friction can be productive as "clashes between different temporal orientations open up a space for strategic manoeuvring and social negotiation" (Ringel, 2013, p. 32) and new entry points for innovative planning approaches and activity (Baeten, 2012). Equally, these frictions can contribute to increased precarity for particular stakeholders.

As places where different temporalities of action come into friction (Sassen, 2000) with temporalities of inaction or "'negative' temporalities of delay or failure", unpacking the evolutionary politics of city-spaces helps complicate understandings of the past, present and future (Abram & Weszkalnys, 2011, p. 14). While generally produced by the structural rupturing of linear time, stasis also emerges from divergent framings of time by different actors. This may occur for example through planning appeals by particular stakeholders, whose sole purpose is to privilege the past over any future development. Ricoeur (1984, p. 9) argues that "a person's present disposition can only be made sense of in terms of a future and past, so the present becomes an expanded field". But the present is distended differentially by different agents, with varying levels of power, producing a temporal flexibility that is culturally "arbitrary and at the same time (politically) significant" (Ringel, 2013, p. 31).

Time variably delineated by different actors, can become strategically deployed to achieve particular objectives. Socio-politically and politico-economically defined, these tend to focus on the immediate present or long-term future with the recent past and immediate future neglected (Guyer, 2007), shaping subjective experience. While Lefebvre and others suggest that control *of*, and *over*, time is the preserve of select and powerful interests, a "dualistic opposition between 'policy makers' and 'community'" (Lombard, 2013, p. 7), we contend that contemporary urban governance demands a more granular understanding of the role and power of urban agents. Through the lens of temporal politics, binary understandings of power in the neoliberal city as state/market versus people/community are opened to challenge. We suggest that in addition to traditional, class-based power differentials that produce particular temporalities and frame certain formal agents as elites, power based on symbolic capital defined through time-space (Bourdieu, 1984) is producing new forms of urban elites. These new elites inherit their honoured status (as descendants of the rebellion leaders, in our case) through temporal connections to particular places (Lilja, Baaz, & Vinthagen, 2015). Such elites come into tension with traditional powerholders – planners, developers, financiers – to reflect and contest, "how the future, a powerful experiential dimension in times of change, is differently imagined by a whole repertoire of conflicting temporal narratives" (Ringel, 2013, p. 34). In doing so, it becomes clear that while various actors play with time in different ways, some temporalities are *beyond the control* of traditional urban development actors with major implications for governance and the likely success or failure of particular plans. Those wielding social power may have greater control of temporality – however framed – but they do not necessarily have full control, highlighting the impotence of governance processes over both time and urban temporalities. As new elites with power legitimised by their personal and political pasts enter the urban arena, this becomes even more complex (Lilja et al., 2015). In the next section, we introduce our empirical study focusing on the planning, governance and heritage context of proposed redevelopment in Moore Street and its market before moving to an in-depth discussion of how temporal politics are being produced through space and time in Dublin and with what implications.

Moore street and its environs in context

Planning and development context

As a city, Dublin has been subject to a number of boom-bust development cycles over the past thirty years. Emerging from significant urban decline following the recession of the late 1970s and early 1980s, a fundamental step-change in planning occurred in the mid-1980s with the introduction of a wide-ranging Urban Renewal Act (1986), target-ing specific sections of inner city Dublin as well as other cities for redevelopment. Central government renewal programmes aimed to address physical dereliction, combat social disadvantage and promote economic development in key parts of the city. They adopted particular entrepreneurial approaches to redevelopment and governance of regeneration areas, including the establishment of specific development agencies out-side the remit of the traditional planning system (Moore, 2008). The initiatives evolved from "pump priming" private sector investment in some areas in the 1980s, to a more integrated area-based planning approach in the 1990s, driven primarily by fiscal incentives.

The first attempt at redevelopment in Moore Street emerged as part of the 1999 Integrated Area Plan (IAP) for O'Connell Street. The need to upgrade Moore Street was identified but the importance of its retention as a market area was also acknowledged. Initial plans included the partial-roofing of the street and market but a decision by Dublin City Council (DCC) in September 2002 to close the Smithfield wholesale fruit and vegetable market, where Moore Street traders procured their produce, put the viability of any upgrading into question. In 1999, permission had also been granted for a major commercial development, Millennium Mall, covering a very large site from O'Connell Street to Moore Street threatening the continuation of the market.

In 2001 with no sign of development commencing, Dublin City Council launched a Compulsory Purchase Order for the site, which although appealed, was confirmed in 2007. Property and investment group Chartered Land acquired the site adding it to a significant landbank of over 5.5 acres that they had been amassing over many years. Planning permission for a €1billion commercial development, the "Park in the Sky", comprising 100 high-end retailers and 250 apartments was sought in 2008. The proposal suggested the enhancement of Moore Street and the retention of the market, albeit rebranded as a "foodie heaven" (Irish Independent, 22 April 2008). However, this proposal – coming fourteen years after the first proposals by Green Party Councillor Ciarán Cuffe to promote a European style open-air market in the area – was greeted sceptically by traders who had become increasingly disillusioned with their place on the street and yet more plans. Although the arrival of new migrant entrepreneurs on to the street in the late 1990s enabled the marketscape to retain an element of vibrancy, increasing negative media attention linked to long-term disinvestment caused long-

term traders grave concern: "people come from all over the world to see Moore Street
... this is not what they expect to see, it's derelict and the council has let it get this way"
(Catherine Kennedy, *Irish Times*, 17 March 2015).

Traders highlighted systemic failure by the local planning department: "every few
years there's some sort of new plan, but I think the council will be happier when we're
just gone." (Marie Cullen, *Irish Times*, 17 March 2015). This precarity was intensified by
the stasis that accompanied the Great Recession from 2008, creating what Wallace
(2015) describes as a "limbo-land".

Heritage and development context

Complicating matters has been the heritage significance ascribed to Moore Street in
recent years. In the lead-up to the centenary of the 1916 Easter Rising, the street became
the focal point for national narratives of nation-building. While the insurrection lasted
only days before being quashed, the leaders of the rebellion surrendered from house no.
16 on Moore Street. As plans for national centenary commemorations were being
made, the area became central to a discourse of "nation-building" by campaign groups
and nationalist politicians, conferring an historical significance on this space that had
previously been relatively ignored.

Initially ignited in the late 1990s by the National Graves Association "Save no. 16
Moore Street" campaign, this gained momentum at the turn of the millennium. In the
spring of 2003, city councillors voted to preserve the building. In the context of the
emerging development plans, protracted engagements between the city manager, building owners and planning authorities resulted in agreement to declare the building a
national monument, allowing redevelopment to begin around it. A preservation order
was placed on house no. 16, and a wider block comprising houses no. 14–17 was
declared a national monument.

Located within the footprint of the proposed "Park in the Sky" project, heritage
campaigners were outraged when planning permission was granted and appealed the
decision. In 2009, the planning appeals board's assessment called for major changes to
the proposed development. However by that time – and in the context of a deepening
crisis that highlighted the centrality of time to the capital accumulation process –
development had stalled across the city. The Great Recession of 2008 had catastrophic
impacts in Ireland (see Heffernan, McHale, & Moore-Cherry, 2017) including negative
urban implications (O'Callaghan et al., 2015; O'Callaghan & Lawton, 2016). The
planned commercial redevelopment for Moore Street was one of the high-profile
casualties as the developers involved were declared bankrupt, and the development
halted. Institutionally, change was also afoot. During 2009, Ireland's central government
established the NAMA (National Asset Management Agency). Effectively operating as a
bad bank, NAMA acquired property loans from five main banks in a bid to stem a
systemic banking crisis. Loans secured on properties in the Moore Street area were
transferred, and heritage campaigners argued that this provided an opportunity for the
state to take control of this "national heritage" site. Central to their argument was the
timeliness of this state acquisition in relation to the impending centenary and the
previous "failures" of central government to adequately recognise the significance of
this urban quarter.

However NAMA, as a commercial entity with responsibility for maximising returns to the taxpayer, opted to provide €10million in development funding to the developers to facilitate a new planning application and jumpstart redevelopment. This new project proposed the demolition of the "1916 national monument" as well as the open-air marketplace. Despite widespread political opposition and growing support for the campaign, the Irish central government refused to intervene. Only in April 2015 – exactly one year before the centennial celebrations – did it actively get involved buying numbers 14–17 Moore Street from NAMA. In November 2015, the Office of Public Works began restoration work on the buildings for a commemmorative centre but activity was halted in January 2016, as a legal challenge by campaigners was brought to the High Court of Ireland. This was successful and the judge, Mr Justice Max Barrett, ruled that a wider "battlefield site" should be recognised and given state protection from private development and that all ongoing work should cease (see Figure 1). Given the potential wider implications for planning and development law, the Irish government appealed the ruling and the case has yet to be heard (December 2017). While the "battlefield site" designation was heralded as a major success for the campaigners, it highlighted the failure yet again to generate a coherent spatial imagination for the street, complicated the governance of the street through the judicial process, and sustained the precarity of livelihood-building

Experiencing time on moore street

Drawing upon our earlier discussions of the multiple temporalities of the city and on the evolution of Moore Street, in this section we explore the temporal politics that have played out and given rise to its contested place in the city. A plurality of temporalities coexist and intersect in this space; long-standing local and everyday rhythms that ground and shape the marketplace contrast with other temporal strategies that relate to Moore Street's redevelopment and directly affect the marketscape's day-to-day functioning and spatial imaginations of the street. Moreover, the idea that time is experienced differently urges us to pay attention to *who* experiences it as, "time is lived at the intersection of a range of social differences that include class, gender, race, immigration status, and labour" (Sharma, 2017, p. 194). This is very relevant in the case of Moore Street where an attention to these dynamics can enable a better understanding of the complex temporal politics at play.

Livelihoods and living informal politics through time

Up until the 1960s, Moore Street had assumed a core role in food provisioning of the inner-city and a visit to the marketplace was a key part of many urban resident's daily routine. Daily, weekly and seasonal rhythms of neighborhood food access were all shaped through Moore Street and other similar public markets in the city. Interviews with long-term traders, all of whom were women, and others who had personal experiences of childhood trips to the market as part of their daily familial routine, infused their descriptions of these encounters with powerful emotions, attachments, and memories. One long-term trader in her 60s explained:

Dubliners love the market, love Moore Street. They love the idea of coming down because they all have stories to tell from when they were kids, coming with their mum. They'd be made to come down and carry all the shopping home. They all have stories to tell, the Dubliners (Interview, 26 September 2016).

More recently, temporalities of diaspora and migration have been produced through Moore Street as its powerful symbolism continues to bind distant people to their former hometown. For many Dubliners, those currently living in the city as well as those who have now moved away from it, Moore Street invokes memories of a north inner-city Dublin based on a strong sense of community belonging, a place of good banter and "craic" (fun), and a moral economy where people took care of each other. Above all, this was epitomised by the market, and by the traders who filled it with life and character (Figure 2). A number of participants commented, perhaps in a mobilisation of nostalgia for particular ends (Meeus et al., 2016), upon Moore Street's role in connecting Irish migrants back to Dublin and to memories of home. As one long-term resident related:

That's what people think of when they think of Dublin, like, when they go abroad. I've got relatives that left when they were young kids and emigrated off to a great grand aunt in America. And like, she's really passionate about Moore Street...She says her vivid memory is going to Moore Street and getting the shopping, and vegetables and the traders and the accents and everything. Yeah, so vibrant. So it's something we really can't afford to lose, you know? (Interview, 12 August 2016).

Yet from the 1960s, as inner-city residential communities became increasingly dispersed through large-scale relocations to a nascent suburban Dublin, new formats of shopping and food provisioning, namely the spread of local grocery stores, began to gain ground. This dramatically changed the food retail landscape and the rhythms of households' food provisioning routines; by substantially shortening the shopping trip, it became less of a key social and family "event" in daily life.

Yet just as Moore Street's customer base was shrinking, a dependency on the market as a source of livelihoods was beginning to grow amongst working class local residents as traditional industries declined in the inner-city and businesses relocated to the

Figure 2. Moore street traders and customers in the mid-1980s.

suburbs. The gendered dimensions of this economic restructuring involved a massive displacement of men from industrial labour-based livelihoods. Given its very easy access, flexibility and minimal start-up costs, market trading became an important livelihood venture for women in households needing an alternative source of income. It became common for the growing ranks of women traders to "triple up" their baby prams as a cart to carry their wares and to use as a mobile market stall. Early in the morning, women from the inner-city would walk the kilometre from Smithfield's wholesale market, where they purchased cheap fruit and vegetables, to sell them from their prams to other locals in the Moore Street market. Around this time, these converted prams also became a powerful symbol of women traders' resistance and resilience during crackdowns on street trading in the 1980s. Through protest and political supporters who maintained that traders had a constitutional right to earn a living, the traders managed to obtain some concessions. The market has remained a symbol of the city to this day but has come under increased pressure as property-led urban redevelopment agendas increasingly encroach on the "marginal" spaces of the city.

While the market carries on, the traditional trade as a whole has been retreating. A big decline in trader numbers was brought about by the closure of the Smithfield wholesale market in 2002, the main supplier for Moore Street traders, as noted earlier. This was then compounded by an increase in stall license fees by Dublin City Council. A trader count in November 2016 revealed a total of 25 market traders currently remain (20 selling fruit and veg, 3 flower sellers, and 2 fish sellers sharing one stall), half the number from a decade ago. In addition to a drop in overall numbers, interviewees described how the rhythms of trader's livelihood patterns on the street have changed in the last couple of decades, with a marked reduction in frequency and regularity:

> A number of those stalls wouldn't be as active now. Before, people would have been there all the time. And if for some reason – whatever, recreation, holidays, or sickness – people couldn't be there, there was always a member of the family or extended family to stand in to keep the stalls running, whereas today I think we see only 3 or 4 stalls like this left. There were always fish stalls on Moore Street and now I think they only operate a few days a week. And even in the last two years, they diminished from 3 to now I think there's only 2 that are there (Interview with long-term resident, 28 August 2016).

Temporal connections to the past features strongly for many of these traders who have decided to stay put; their familial histories and connections act as a form of informal resistance (Lilja et al., 2015) to the imposition of a dominant spatial narrative by other stakeholders. For this group, their livelihood represents more than a material existence and is about identity and a way of life. Being a trader on Moore Street is viewed as an important part of their genealogy; a few market traders explained that they could trace back three or even four generations of trading on Moore Street. This formed their key rationale for remaining despite the increasing challenges with which they are faced. The market physically connects them to past generations of their families and the long-standing symbolic significance of street trading allows them to assert their place and sense of belonging in a rapidly transforming city (Watson & Wells, 2005). While their discourses are shaped with a strong sense of past tradition, traders also recognize and embrace change to ensure a future for the marketplace, highlighting the complex ways in which the past, present and future are linked (Crang, 2001).

A dramatic social transformation of the street has also occurred since the late 1990s, with the arrival of migrant traders and shopkeepers, particularly of Eastern European, African, South and Southeast Asian, and Chinese origin. Offering a variety of specialty ethnic products and retail services catering to Dublin's migrant communities, these traders are regarded by some of the traditional traders and local residents as having infused the marketscape with a new vibrancy and helped sustain it in the face of external pressures. As a national politician described:

> We're seeing the market reflect the changing demographics of the north inner-city. Forty percent of those who live in the north inner-city weren't born in Ireland, and we're seeing that in the market; we're seeing African, Eastern European, and Asian retail activity, not so much in the market, in the street market, but in the shops around the market. I think it's a good development and I think it needs support and encouragement to assist particularly recent immigrants to get a foothold on the retail ladder and to contribute to the dynamic of the place (Interview, 19 July 2016).

Nevertheless, both this more recent and the long-standing traditional traders on Moore Street are under serious threat as temporal strategies tied into the redevelopment agenda clash with the rhythms of livelihood building (Sassen, 2000). One visible manifestation of this is a highly transient marketscape with new shops opening and closing within a relatively short time-frame. This transience is often not borne out of choice (Dublin Enquirer, 2/11/2016). Those renting shops along Moore Street owned by the developer must accept to operate their businesses under very short-term lease arrangements (3 month and even 30-day contracts) so that the premises can be quickly vacated whenever the green light arrives. This, in conjunction with sudden and recurrent rent increases, has encouraged a high turnover of businesses on the street. As a local resident noted:

> They tend to be nearly like pop-up shops anyway. I don't think they even have to be thrown out because they're on a short-term lease. They tend to be very transient kind of set-ups. (Interview 12 August 2016).

When we asked one shopkeeper who moved to Ireland from China seven and a half years ago what he thought about these changes, he responded:

> There used to be an array of shops, but these are all gone now. For Ireland, I think it is a positive change, as there is a plan for a large shopping centre to be erected at this street. However, for myself, my shop may not exist here anymore. So it is not a good thing for me. But I cannot do anything. It is the power from the governments (Interview, 16 April 2016).

What is evident – whether traditional traders or migrant enterprises – is that the temporal cycles of livelihoods in the marketplace are being gradually squeezed under the tacit promotion of the Dublin City Council. Traditional street traders are still permitted to hand down permits within trader households, which would appear to support the continuity of livelihood practices over the longer-term. However, Dublin City Council have prohibited the distribution of new stall licenses, effectively capping the total number of traders to what are currently left, hindering any expansion. This approach reflects an implicit "playing with time." New regulations are central to the increasing inability of the market to survive, raising important questions around who defines urban policy success or failure. Traders explained how the younger generation are no longer interested in carrying on this livelihood tradition. While no upper time limit has been placed, the longer-term cycles of livelihood renewal over the life course are likely to end in the near future when the current traders retire. This passive control over time by officials via the approach of "waiting it out" (Wallace, 2015), has been compounded by the much more explicit temporal tactic of issuing short-term leases to migrant traders to ensure that redevelopment can proceed at any time. The experience of this informal politics of time has reinforced the precarity of already vulnerable and marginal social groups and spaces within the city. This has been buffered by the more formal politics of planning which has produced other, more diverse, temporalities that have been strategically deployed by, and are constructive of, new elites on the street.

Living in limbo: disinvestment and broken promises

Despite repeated promises and plans for upgrading, traders in Moore Street are frustrated by Dublin City Council's failure to make good on promises over many decades of upgrading and infrastructural improvements, only the most superficial of which have materialised. While the local authority declares its support for the market in principle, in practice traders have been without key basic infrastructure like water supply, toilet and sanitation facilities, adequate lighting and proper waste disposal. Traders note that they have been waiting for street lighting for almost twenty years, while they were promised a water supply as much as thirty years ago. One trader voiced her irritation as follows:

> ...everything is about, 'Do you want to upgrade the market?' You look around, the market is not great. We don't even have proper street lights... and I was told last year – and this is 17 years down the road that I've been asking for street lighting in Moore Street – 'Oh, we don't have it. We've no money available. There's no funds available.' But yet, you pick up the paper and you look, 'Grafton Street [an upmarket shopping street in Dublin's city centre] is getting granite footpath and paving'. So many million, 17 million or whatever. That's for people to walk on. And we're not allowed to get proper street lighting, from the same people that's giving these people...So what are we, the poor relations? It's as if Moore Street doesn't exist. But what they're forgetting is that Moore Street is the heart and soul of Dublin, of Ireland. And we're just being ignored (Interview 23 August 2016).

Many traders have interpreted these extreme time-lags and the persistent failure of Dublin City Council to invest in the market as a clear message that the overarching strategy – within the context of a deepening neoliberalism of urban policy in Dublin (MacLaran & Kelly, 2014) – is to force the market to disappear. As one trader shared:

> ...they haven't invested [in us] – I mean, they borrowed us a couple of tables. Is that supposed to enhance the street? I don't think so. I think if we got proper street lighting, proper pavement, proper canopies and whatever. They [the tables] look nice but they could be well, well upgraded. It's not as if they're spending hundreds and thousands on it. They're not. So it's just a camouflage act to keep us sweet. You know, 'Give them a couple of tables and let them get on with it. Eventually, we'll get rid of them.' Eventually, they're hoping that we'll get disheartened and we'll walk away from it. But I've no intention of

walking away from it. I'm here to stay. And hopefully one of my children will take it over. Because I'm not going to let the council or the government beat me just so they can do what they want to (Interview 23 August 2016).

When viewed alongside the closure of the wholesale markets and lack of any defined livelihood support for traders, this could be read as a form of passive revanchism of the state (Lee, 2013). The results of this approach become discernible over the long-term through the increasing deterioration of market conditions and the increasingly difficult experiences of living and operating in an environment where time is being played with by powerful stakeholders to set the frame for wholesale change.

In early 2015, as part of a broader strategy to rebrand and regenerate the few remaining markets within the city as a whole, Dublin City Council announced a new plan to redesign Moore Street market. This was to include a new layout, bollards to demarcate trading areas and prevent "sprawling" (despite our evidence suggesting shrinkage has been the dominant trend), a new mandatory code of practice for traders, increased fees, and stringent controls to prevent "anti-social behaviour" and "black-market sales". In light of historical neglect by the local authority, many traders interpreted this as the latest attempt to displace them from the street within the context of imminent construction of the new commercial redevelopment. Dublin City Council's ambitions for the city markets were admirable, but within such a politically charged environment they were unlikely to be realised. While much media interest was focused on the 1916 heritage campaign, scant notice was paid to the plight of Moore Street's traders and businesses, or their voice in debates about the future of the street.

While the planning process brings the current and future states of the street into conversation in particular ways, this occurs within a highly structured and linear approach to time that marks and views time and space in particular technocratic ways through development plan timespans, prescribed timelines for consultation and other similar actions (Raco, 2008). Although this rational approach to and rationalization of time is highly at odds with the temporalities of those who depend on the street for their livelihoods, there has been a co-existence or at least a worked negotiation of these temporalities in operation for many decades. While far from ideal, this "balance" has been challenged and made more complex in recent years through the insertion of the heritage campaign into this urban space, bringing pasts, presents and futures into interaction in different but highly contested ways (Crang, 2001).

Temporal (in)flexibility and challenges of governance?

Attempts by formal stakeholders such as planners and policymakers to control time-frames to achieve particular goals is not unusual in the contemporary city and indeed is part of the cycle of urban governance. One of the desired outcomes is usually the assertion by elite stakeholders of a dominant future spatial imagination of particular sites and neighbourhoods (see Kaika & Ruggiero, 2015, 2016).Yet Moore Street appears to be anomalous as it has become impossible to articulate any agreed spatial imagination for the street and market, representing a failure of not just policy but also urban governance. We argue that this can be explained through the different forms of temporal reasoning being employed to contest not just the future redevelopment of

Moore Street but also its pasts and presents. Public planning bodies and urban developers are no longer the only actors engaged in the process of projecting the future. Over the last decade this has been overlain by a highly complex and multi-layered struggle over the rightful heirs to its history (Lilja et al., 2015) and the production of alternative, competing futures for the district.

These struggles are the result of new and powerful voices associated with a diversity of heritage campaign groups. We consider the heritage campaigners as an alternative set of elite actors – that acquire their power through defined temporalities (Bourdieu, 1984) – alongside the "traditional" set of redevelopment elites in the neoliberal city, comprised of local government, urban planners and private developers. The significant presence and influence of these other elites in setting the terms for whether, when, and how Moore Street's future redevelopment occurs serves to complicate the standard dichotomy of urban redevelopment between the state/market and the local community. For the heritage groups, "the past" is selectively constructed around [their] particular group's narratives and memories,' (Bastian, 2014, p. 148) and particular aspects of Ireland's history are privileged to underpin their struggles (Figure 3). Resistance through the deliberate stalling or halting of redevelopment has occurred with little regard for the potential consequences of these actions on rhythms of urban livelihoods. However, ironically, this deployment of particular temporal framings that threatens the market is also producing a stasis that may partly be sustaining its continuance in the face of proposed commercial redevelopment. This complexity is compounded by evidence that suggests the success of campaign groups in achieving stoppage against the will of more formal power agents (local authority, central government, developers) was made possible by a history of failed planning and development in the area over many decades.

However, these new elites are far from a united group, and within the campaign itself, there has been much fragmentation. There are at least four different groups campaigning around similar themes, but fractured through internal disagreements

Figure 3. "Heritage" graffiti in Moore Street, Dublin (2016).

and conflicts. These campaigners were discussed humorously by one long-term local resident:

> It's a bit like that scene in Monty Python, you know, the Judean People's Front and The People's Front of Judea! Lots of people involved in this campaign have invested themselves in the garb of their ancestors, as if they themselves are the signatories of the Proclamation in 1916 which, to me, hugely smacks of hubris and ego. And, you know, I think be your own hero. Don't trade on your ancestor's reputation to the exclusion of all else, which I think has happened to a large extent. (Interview, 12 August 2016).

The temporal reasonings of campaign groups are also similarly splintered. For example, one group, the original Save 16 Moore Street group wanted the preservation of a small number of the buildings on Moore Street most directly involved in the Easter Rising. Another group, the 1916 Relatives Association, is demanding a much greater concession in terms of the "heritage" of the street and is aligned with more left wing republican politics. Indeed, the failure by the centre-right governing party (Fine Gael) to act more quickly and decisively to appropriately mark the centenary only fuelled the highly publicised struggles that ultimately ended with the High Court case in March 2016 that declared a large part of Moore Street and its environs as a "battlefield site" (Figure 1). Until the appeal is heard, no progress in terms of historic preservation, market upgrading, or commercial redevelopment can be made.

Where does the Moore Street market and those whose livelihoods depend on it fit into these arguments? While a number of campaigners we interviewed confirmed that they personally regarded the marketplace as of historical value – some even arguing that it should be part of the campaigners' heritage narrative – on the whole, the marketscape and those whose livelihoods are based there, are voiceless and invisible in these campaigns. The narrowed fixity upon a particular historical timeframe and precise location (i.e. a particular set of buildings) has overlooked a much broader and potentially more powerful approach to historical preservation that could include a temporally extended and spatially encompassing view of Moore Street. Drawing on Guyer (2007), this highlights the manner in which a focus on the distant past or long-term future can over-ride the recent past and immediate future, with particularly acute consequences for those for whom these timeframes shape their lived experience of the city.

Indeed one of the key failures in this story has been the inability of the heritage activists to build the support of the traders into their campaign and thus broaden their support. Their temporal inflexibility and lack of inclusivity has not been lost on Moore Street's traders, some of whom feel alienated from a movement that could have fostered solidarity via a more inclusive alliance. This might have been possible if the campaigners' arguments were clearly linked to an anti-gentrification or working class social history agenda, rather than historical rhetoric based on nationalist credentials. One long-term trader on the street captured the essence of the predicament:

> the market will probably be dead... Unless something is done. Honest to God, unless there's something done ...We're not even coming into it. It's all about this building and everything else. They don't even know what they're arguing over. It's just one group trying to get at another group. (Interview, 26 August 2016).

This clearly reflects "how (culturally) arbitrary and at the same time (politically) significant the application of pasts is in social negotiations of identity and belonging" in the city (Ringel, 2013, p. 31). As one long-term resident noted:

> ...there's some very good people on the campaign but there are also people who walk down to the market, they bring politicians down, they don't talk to traders, they swan past them. They walk past the *living heritage* of the market to put their hands on cold brick, which seems to me to be hugely short-sighted. To leave the traders out is to leave everything out (Interview, 12 August 2016).

If Moore Street's traditional market traders have been marginalised from the campaign, the numerous migrant-run shops and businesses on the street have been even more so. While migrant owned and operated enterprises have enabled the resilience of the market street, they have been completely absent from the heritage campaigns' vision of the street and are perceived as an entirely detached element from the campaign. When we asked one such shopkeeper whether he had been consulted by the campaigners, he responded, "No, because we are aliens and not locals...We are foreigners and we haven't been staying here long enough, so they were not targeting us and hearing our ideas about Moore Street" (Interview, 16 April, 2016). Yet, any change happening in and to Moore Street will have a direct and strong impact on this group of people.

Heritage campaigners have successfully affected "planning time" by pursuing a series of appeals, to the effect that the time gap between these plans and their realisation becomes an enduring state of affairs. But for those who experience this planning time as an ever-distended present, it is a frustrating condition . One trader voiced her opinion on the impact of redevelopment time lags as a result of the heritage campaign:

> I'll tell you what.... I don't think they're helping us. They have their own ideas and what they're doing it for. Let me tell you, waiting 100 years and still getting nothing done for the 100 year anniversary was a disgrace. And the only people to blame are the people that stopped everything, put a stop to everything...As far as I'm concerned, they didn't do Moore Street any favours – and they certainly didn't do Michael Collins [rebel leader] and all his comrades any favours. Because there's that house over there and it's still like a derelict site. It could have had a beautiful museum, something, something. Anything (Interview, 23 August 2016).

Traders and businesses on Moore Street who continually experience planning time as a never-ending series of delays, have become ambivalent or even resentful about both Dublin City Council and the heritage campaigns alike. For these individuals, Moore Street's limbo condition has gone on so long that enough is enough; the local authority is perceived as failing the traders through the continued revision and lack of implementation of plans and policies. An intervention by the developer would be welcomed even despite the uncertainty and risk that redevelopment poses to the market's existence.

Conclusion

Cities are produced in and through time and space (Crang, 2001; Harvey, 1985). They are assemblages constituted by "travels and transfers, political struggles, relational connections, and territorial fixities/mobilities" but within a context of multi-dimensional and superimposed cyclical, alternating and linear rhythms and diverse temporalities (McCann & Ward, 2011, p. xv). By adopting a focus on how time is experienced in cities, or the temporal politics of the urban environment, we have adopted a more grounded and nuanced approach to unpacking the evolutionary politics of particular city-spaces and the politics – both formal and informal -of planning. In this paper, we have argued that cities are shaped by temporalities. The city is an arena where different pasts, presents and futures come into friction with one another, highlighted through our case study of Moore Street and its market. These are places where repeated plans for the future have been developed but left unimplemented. While heritage campaigners actively managed to block any redevelopment on the street using powerful, compelling narratives of historic preservation and nation-building, the 1916 centenary has now passed with no material commemoration in this significant location.

On the surface, Moore Street accords to what Wallace (2015) defines as a "limbo-land", where time has ground to a halt. Yet beneath the surface, time has raced on. Financiers and bankers have experienced time very differently in this space; significant mobility has been occurring in terms of financial transactions and changes to land ownership, with at least three different owners identifiable in the last decade. This transfer of ownership of buildings that have been physically fixed in space over long periods of time, occurs within the context of increasingly mobile international capital flows facilitated by a deepening form of neoliberal urban governance. This temporality neither registers for those who experience the daily rhythm of the city through market trading nor those who campaign to privilege a different temporality.

Our case study raises important questions about the governance of temporalities. Documents such as development plans are seen as a tool to control time and planning, to create fixity and certainty, with writers such as Lefebvre (in Lefebvre, 1996) arguing that planners are the New Masters who control time. We argue that it is not so simple. New elites – such as the heritage campaigners in our study – are constructed through time and space and attempt to control time through strategies such as planning appeals and legal interventions, deploying different temporalities strategically to do so (Ringel, 2013). What does this mean for urban governance and the potential success or failure of any particular policy or plan? In an entrepreneurial governance framework where "streamlining" or "fast-tracking" of permissions and development is key, under-estimating the power of these new elites could be detrimental.

The story of Moore Street clearly highlights the changing nature of power relations in the city and the importance of a diverse set of actors in defining which temporalities come to bear most strongly on the future city. The practical outcome of the legal challenge brought by heritage campaigners to the redevelopment plans has been the judicialisation of planning. Indeed, a detailed analysis of activity on the street since the 2000s, from an initial compulsory purchase order through to the current High Court battlefield designation and its future appeal, highlights the growing power of the legal system in shaping future urban environments. This has created uncertainty, shown the potential impotence of traditional planning mechanisms and the restrictions of thinking about the city through a linear set of technical-rational timeframes.

Determining policy success and failure in such rapidly changing contexts is therefore challenging. Whether the recent past and future of Moore Street will be seen as a "success" or "failure" depends on the temporal framings we privilege. Whether we frame our interpretation through the eyes of the traders, campaigners, central government, local authority planners, or developers in the short, medium, or long-term will provide a very different reading of the street. What we can see objectively is that the long-term functional resilience of this marketplace within a highly challenging environment could be interpreted as success. Yet simultaneously, the short-term physical dereliction and failure to implement plans might be read as failure. Is the ability of the heritage campaigners to stall development in order to preserve particular buildings or achieve battlefield designation a success? Or was it made possible by a failure of proper planning and development? What is clear is that the formal and informal politics associated with how time is experienced in and through particular places is fraught with contestation and provides significant challenges for urban governance.

Acknowledgements

This research was funded by an Irish Research Council New Horizons research grant. The authors are grateful to Zhao Zhang and Niall Traynor who assisted with data collection, and to members of the Critical Political Economy cluster at University College Dublin, to Clare Mouat, John Lauermann, Cristina Temenos, and the three anonymous reviewers for their constructive and insightful comments on an earlier draft of the paper.

Disclosure statement

No potential conflict of interest was reported by the authors.

Funding

This work was supported by the Irish Research Council [grant number: 48622]

ORCID

Niamh Moore-Cherry ⓘ http://orcid.org/0000-0003-0372-8809
Christine Bonnin ⓘ http://orcid.org/0000-0001-5391-5050

References

Abram, Simone, & Weszkalnys, Gina. (2011). Introduction: Anthropologies of planning—Temporality, imagination, and ethnography. *Focaal: Journal of Global and Historical Anthropology, 61*, 3–18.
Baeten, Guy. (2012). Neoliberal planning: Does it really exist? In Tuna Taan-Kok & Guy Baeten (Eds), *Contradictions of neoliberal planning* (pp. 205–211). Netherlands: Springer.
Bastian, Michelle. (2014). Time and community: A scoping study. *Time & Society, 23*(2), 137–166.
Bishop, Peter, & Williams, Lesley. (2012). *The temporary city*. London: Routledge.
Bourdieu, Pierre. (1984). *Distinction: A social critique of the judgement of taste*. Cambridge, MA: Harvard University Press.

Castells, Manuel. (1997). *The information age: Economy, society and culture. Vol. 2, The power of identity*. London: Blackwell.

Crang, Mike. (2001). Rhythms of the city: Temporalised space and motion. In Jon May & Nigel Thrift (Eds), *Timespace: Geographies of temporality* (pp. 187–207). London: Routledge.

Degen, Monica. (2017). Urban regeneration and resistance: Foregrounding time and experience. *Space and Culture, 20*(2), 141–155.

Dines, Nick. (2007). *The experience of diversity in an era of urban regeneration: The case of Queens Market, East London*. Milan, Italy: Fondazione Eni Enrico Mattei (FEEM).

Dodgshon, Robert. (2008). Geography's place in time. *Geografiska Annaler: Series B, Human Geography, 90*(1), 1–15.

Gonzalez, Sara, & Dawson, Gloria. (2015). *Traditional Markets under threat: Why it's happening and what traders and customers can do*. Report. Leeds.

Gonzalez, Sara, & Waley, Paul. (2013). Traditional retail markets: The new gentrification frontier? *Antipode, 45*(4), 965–983.

Guyer, Jane. (2007). Prophecy and the near future: Thoughts on macroeconomic, evangelical, and punctuated time. *American Ethnologist, 34*(3), 409–421.

Harms, Erik. (2013). Eviction time in the new Saigon: Temporalities of displacement in the rubble of development. *Cultural Anthropology, 28*(2), 344–368.

Harvey, David. (1985). *The urbanization of capital: Studies in the history and theory of urbanization*. Oxford: Blackwell.

Heffernan, Emma, McHale, John, & Moore-Cherry, Niamh. (2017). *Debating Austerity in Ireland: Crisis, experience and recovery*. Dublin: Royal Irish Academy.

Henderson, Steven, Bowlby, Sophie, & Raco, Mike. (2007). Refashioning local government and inner-city regeneration: The Salford experience. *Urban Studies, 44*(8), 1441–1463.

Jessop, Bob. (2006). Spatial fixes, temporal fixes, and spatio-temporal fixes. In Noel Castree & Derek Gregory (Eds), *David Harvey: A critical reader* (pp. 142–166). Blackwell: Oxford.

Kaika, Maria, & Ruggiero, Luca. (2015). Class meets land: The social mobilization of land as catalyst for urban change. *Antipode, 47*(3), 708–729.

Kaika, Maria, & Ruggiero, Luca. (2016). Land financialization as a 'lived' process: The transformation of Milan's Bicocca by Pirelli. *European Urban and Regional Studies, 23*(1), 3–22.

Lauermann, John. (2016). Temporary projects, durable outcomes: Urban development through failed Olympic bids? *Urban Studies, 53*(9), 1885–1901.

Lee, Peter. (2013). Housing Market Renewal: Evidence of revanchism or a response to 'passive revanchism' supporting 'citizenship of place'? *Housing Studies, 28*(8), 1117–1132.

Lefebvre, Henri. (1996). *Writings on Cities*. Oxford: Blackwell. Translated and edited by Eleonore Kofman and Elizabeth Lebas.

Lefebvre, Henri. (2014). *Rhythmanalysis: Space, time, and everyday life, Translated and edited by Stuart Elden and Gerald Moore*. London: Bloomsbury.

Lilja, Mona, Baaz, Mikael, & Vinthagen, Stellan. (2015). Fighting with and against the time: The Japanese environmental movement's queering of time as resistance. *Journal of Civil Society, 11*(4), 408–423.

Lombard, Melanie. (2013). Struggling, suffering, hoping, waiting: Perceptions of temporality in two informal neighbourhoods in Mexico. *Environment & Planning D: Society and Space, 31*(5), 813–829.

MacLaran, Andrew, & Kelly, Sinead. (2014). *Neoliberal urban policy and the transformation of the city: Reshaping Dublin*. Netherlands: Springer.

McCann, Eugene, & Ward, Kevin. (2011). *Mobile urbanism: Cities and policymaking in the global age*. Minnesota: University of Minnesota Press.

Meeus, Bruno, Tim, Devos, & Seppe, De Blust. (2016). The politics of nostalgia in urban redevelopment projects: The case of antwerp-dam. In Christian Karner & Bernhard Weicht (Eds), *The commonalities of global crises* (pp. 249–269). London: Palgrave Macmillan.

Moore, Niamh. (2008). *Dublin Docklands Reinvented: The post-industrial regeneration of a European city quarter*. Dublin: Four Courts Press.

Moore-Cherry, Niamh. (2017). Beyond art in 'meanwhile spaces': Temporary parks, urban governance and the co-production of urban space. In Monika Murzyn-Kupisz & Jarosław Działek (Eds), *The impact of artists on contemporary urban development in Europe* (pp. 207–224). Netherlands: Springer.

Moore-Cherry, Niamh, Crossa, Veronica, & O'Donnell, Geraldine. (2015). Investigating urban transformations: GIS, map-elicitation and the role of the state in regeneration. *Urban Studies*, *52*(12), 2134–2150.

Moore-Cherry, Niamh, & McCarthy, Linda. (2016). Debating temporary uses for vacant urban sites: Insights for practice from a stakeholder workshop. *Planning Practice & Research*, *31*(3), 347–357.

Munn, Nancy. (1992). The cultural anthropology of time: A critical essay. *Annual Review of Anthropology*, *21*(1), 93–123.

O'Callaghan, Cian, Kelly, Sinead, Boyle, Mark, & Kitchin, Rob. (2015). Topologies and topographics of Ireland's neoliberal crisis. *Space and Polity*, *19*(1), 31–46.

O'Callaghan, Cian, & Lawton, Philip. (2016). Temporary solutions? Vacant space policy and strategies for re-use in Dublin. *Irish Geography*, *48*(1), 69–87.

Peck, Jamie. (2012). Austerity urbanism: American cities under extreme economy. *City*, *16*(6), 626–655.

Raco, Mike, Henderson, Steven, & Bowlby, Sophie. (2008). Changing times, changing places: Urban development and the politics of space-time. *Environment and Planning A*, *40*(11), 2652–2673.

Ricoeur, Paul. (1984). *Time and Narrative – Volume I*. Chicago: The University of Chicago Press.

Ringel, Felix. (2013). Differences in temporal reasoning: Temporal complexity and generational clashes in an East German city. *Focaal*, *66*, 25–35.

Sakizlioglu, Nur Bahar, & Uitermark, Justus. (2014). The symbolic politics of gentrification: The restructuring of stigmatized neighborhoods in Amsterdam and Istanbul. *Environment and Planning A*, *46*(6), 1369–1385.

Santos, Boaventura de Sousa. (2006). *Rise of the global left*. London: Zed books.

Sassen, Saskia. (2000). Spatialities and temporalities of the global: Elements for a Theorization. *Public Culture*, *12*(1), 215–232.

Sharma, Sarah. (2017). Checked Baggage: An afterword for time and globalization. In Paul Huebener, Susie O' Brien, Tony Porter, Liam Stockdale, Yanqiu Rachel Zhou. (Eds), *Time, globalization and human experience: Interdisciplinary explorations* (pp. 191–195). London: Routledge.

Till, Karen, & McArdle, Rachel. (2016). The Improvisional City: Valuing urbanity beyond the chimera of permanence. *Irish Geography*, *48*(1), 37–68.

Wallace, Andrew. (2015). Gentrification Interrupted in Salford, UK: From New Deal to "Limbo-Land" in a contemporary urban periphery. *Antipode*, *47*(2), 517–538.

Watson, Sophie, & Studdert, David. (2006). *Markets as sites for social interaction: Spaces of diversity*. Bristol, UK: Published for the Joseph Rowntree Foundation by Policy Press.

Watson, Sophie, & Wells, Karen. (2005). Spaces of nostalgia: The hollowing out of a London market. *Social & Cultural Geography*, *6*(1), 17–30.

The relational co-production of "success" and "failure," or the politics of anxiety of exporting urban "models" elsewhere

Rachel Bok

ABSTRACT
This paper critically examines the case of the much-vaunted Singapore "model" and its export via the Sino-Singapore Tianjin Eco-city (SSTEC), a megaproject jointly developed by the Singaporean and Chinese states in northeastern China. It revolves around the central question of why, for some Singaporean officials, this export was thought to have "failed" in spite of the model's acclaimed success globally. To address this, the paper historicizes the Singapore model, tracing undercurrents of (geo)political exis-tentialism through Singaporean state meta-narratives that are enacted through thehistorical politics of anxiety and the practitioner politics of anxiety. It argues that categories of policy "suc-cess" and "failure" are relationally co-produced through a politics of anxiety, wherein their stakes are amplified in ways distinctive to small postcolonial city-states. Collectively, the paper emphasizes the enduring significance of (inter)state actors and structures for transnational urban policy mobilization and the limits to assump-tions of post-failure policy learning.

Introduction: the anxieties of small states

In 2015, the postcolonial city-state of Singapore celebrated its 50th year of independence. Amidst the celebratory atmosphere of national commemoration, this anniversary was approached as a routine opportunity for reflection on the city-state's place in the world. The speech given by Prime Minister Lee Hsien Loong during the National Day Rally was no exception:

> We celebrated how we had turned vulnerability into strength ... our journey from third world to first ... [but] big things are happening around us and unless we keep track of events and stay on top of developments, we may be overwhelmed. Singapore is at a turning point. We have just completed 50 successful years. Now we are starting our next 50 years of nationhood. What will this future be? Will Singapore become an ordinary country, with intractable problems, slow or even negative growth; heavy burdens for our children; grid-locked government; unable to act?

> – Government of Singapore (2015)

State meta-narratives of "vulnerability" reveal a longstanding, almost existential fear of failing to participate in globalization, spurred by a more generalized anxiety regarding Singapore's prolonged political sovereignty in the world as a small island-city-state. Small states are constantly subject to political, economic, and military threats (Singh, 1988): in Singapore's case, this has cultivated a "deep security complex" (Dent, 2001) that has historically oriented the scaling of its policy interests toward the external – Southeast Asia, Pacific Asia, the world – to strategically cope with perceived, scripted deficiencies. This positional precarity is couched in an "ideology of survival" (Chan, 1971, p. 154) to remind the populace never to take security and (economic) success for granted, lest failure catch up with the Singapore story. The paper discusses this burden of success, expressed through a preemptive stance against (the prospect of) failure, which extends to the extraterritorial ventures of the Singaporean state in the realm of transnational urban policy mobilization.

Over the last three decades, a diplomatic strategy cultivated by the Singaporean state to negotiate this state of vulnerability has involved the export of its (urban) developmental expertise to countries in the Global South under the rubric of the Singapore "model." The Singapore model is a policy emblem of urban expertise, a post-colony's scripted story of "success," stripped of aberration and abjection, where failure is acknowledged only insofar as it can be narrated as an obstacle that has been vanquished. With China having emerged as a "big power", with game-changing consequences for geopolitical relations globally and within Pacific Asia, the Singaporean state has negotiated this since the 1990s by systematically courting the Chinese as a policy audience. This has been undertaken by exporting the Singapore model to China through customized, intergovernmental channels, resulting in the joint development of new-build cities that are viewed as symbolic of wider diplomatic goals. Lauded by both countries as "flagship cooperation projects between the governments of Singapore and China," there have so far been two developed megaprojects: the Sino-Singapore Suzhou Industrial Park (SSSIP) and the Sino-Singapore Tianjin Eco-city (SSTEC). The SSTEC, in particular, has been extolled by both governments as "a model for sustainable development" (SSTEC, 2019), and viewed in China as a "green urban solution" (Pow & Neo, 2015), the "current 'best practice' model" for its "green leap forward" (Chang et al., 2016). Such are the representations of the SSTEC as one version of "success." If the travels of the Singapore model are intended to exhibit "successful" exports of Singaporean policy expertise to foreign settings, they would appear to have come to fruition in such intergovernmental projects. The significant state investment in the form of sending top-level technocrats overseas, especially on the part of the Singaporean state with its comparatively smaller civil service, makes these projects "exception[s] to the urban norm ... [that] serve as indicator[s] for future urban development" (Caprotti, 2015, p. 36). Herein lies the burden of success for overwrought Singaporean state officials, several of whom viewed the "failure" of the SSTEC as symbolic of the changing power relations between Singapore and China. Such are the travails of the architects of the Singapore model, born of the struggle to reconcile a confidence to make claims of capitalizing on the future, and a fear that reality will not live up to these aspirations, on the other.

This is a paper about transnational urban policy mobilization which revolves around that dialectical tension between assertiveness and apprehension, between the confidence of success and the fear of failure. This is integral to the intrinsic logic of the Singapore model, rooted in the politico-historical context of Singapore's "positionality" (Sheppard, 2002) as a small island-city-state in Southeast Asia. The existing work on urban models has done much to excavate the performative production of success and "best practice." However, as several have argued, success remains but one half of the equation in terms of understanding how differing policy outcomes are enmeshed in broader regulatory land-scapes of power (Clarke, 2012; Jacobs, 2012). Within critical human geography the existing literature on policy mobilization and, specifically, how policy failure is governed, has come to the broad consensus that (the seeming dualism of) success and failure are relationally connected, not least because neither conception "make[s] sense without the other" (McCann and Ward, 2015, p. 828; Temenos & McCann, 2013). Such a relational geography is alert to contingency, contradiction, and unevenness in the processual co-production of policy outcomes across space and scale. The paper contributes to this literature by showing how exactly success and failure are relationally co-produced through a case of South-South policy mobilization between a small state and a big state. Historicizing the Singapore model and its export to China through the SSTEC, the paper argues that success and failure are relationally, inescapably co-produced through a "politics of anxiety."

"Anxiety" is commonly deployed in domains of health and risk, denoting a more generalized state of societal becoming. Wilkinson (2001, p. 17) frames the "condition of anxiety" as "a function of the social predicaments and cultural contradictions in which individuals are made to live out their everyday lives." Critical geographers have con-ceptualized anxiety as a disruptive social practice in the context of food insecurity and consumption (Jackson et al., 2012), apart from general uses of the term that reference emotional states of insecurity, marginality, and ambivalence. This paper takes inspiration from a more sustained approach to anxiety that critically examines state-led develop-mental policymaking models, agendas, and practices, together with the anxiety and abjection experienced by state officials (Doucette, 2020; Sioh, 2010). Doucette (2020), in particular, thinks of anxiety vis-à-vis Anna Tsing's concept of "friction" to foreground the complexity of globalizing connections, viewing anxiety as part of the "mixed emo-tions that accompany, but are often abstracted away from, the discussion of economic models." These social relations, entangled through developmental models, show how "new connections and flows ... also create anxious and awkward entanglements" (Doucette, 2020, p. 3). This paper uses the term "politics of anxiety" to denote the spiraling tension between success and failure as interlocking, mutually reinforcing processes, and the driving force behind the use of the Singapore model as a diplomatic tool of knowledge-sharing. It elaborates on two interrelated realms of this process – the historical politics of anxiety and the practitioner politics of anxiety – that are (geo) politically grounded in Singapore's positionality as a small postcolonial city-state, domi-nated by ethnic Chinese, within the "Malay world" of Southeast Asia.

The empirics are derived from fieldwork undertaken in Singapore and Tianjin from 2013–2015. The paper draws primarily on 15 semi-structured interviews (from a total of 32) with Singaporean state officials who were involved in the SSTEC. I first noticed the heightened awareness of "failure" and "success" during my encounters with Singaporean state officials, many of whom were initially suspicious of the motivations of my research. One defensive official retorted, in response to what I had assumed was a relatively innocuous question regarding the SSTEC's role in industry creation: "Are you just going to say that it's failed, is that the point? You want to see how successful it is and what the problems are?" Much of this secrecy can be traced back to the bilateral aspects of the SSTEC as a Sino-Singaporean project: in state parlance, the SSTEC is a "government-to-government," or "G-G," project whose territorial development is anxiously monitored as demonstrative of the health of Sino-Singaporean relations. The SSTEC is the second G-G project undertaken between Singapore and China; its predecessor, the SSSIP, was considered to have been an embarrassing failure for the Singaporean state. Because of this political primacy and legacy, the SSTEC is (still) regarded as a politically-sensitive topic, especially considering the wider lack of transparency surrounding the extraterritorial ventures of the Singaporean state.

The general impression of researching public officials in Singaporean and Chinese governments is that interviews add little real insight because individuals are hesitant to disclose information that is not already publicly available. While this is not entirely inaccurate, there are ways of circumventing these state-sanctioned silences. Initially, I was told that the "standing policy" of the SSTEC was to decline all interview requests, but once I found a foothold in the "right" networks, snowballing serendipitously propelled me toward the key actors of the project on the Singaporean side, many of whom were prominent members of senior management who had assumed leading roles in the SSTEC because of their background in the Singaporean civil service. I also targeted Singaporean officials who had left the SSTEC, or retired from the civil service altogether, who were more forthcoming. 10 of the 15 Singaporean officials who had participated in the SSTEC (founded in 2008) had also been involved in the SSSIP (founded in 1994), establishing a continuity between both projects that regularly surfaced during interviews. Access to Chinese officials was harder to secure. Hence the paper presents a strong Singaporean positionality that has value in illuminating why the SSTEC was considered to have "failed" for some Singaporean state officials, contextualized within broader historical (geo)political predicaments faced by this small city-state, contrary to public Chinese and international perceptions of undiluted success.

The paper proceeds in four parts. First, it reviews critical geographical scholarship on how policy failure is governed, discussing the difficulty of defining "failure." It complements this with insights from political geographic work on how nation-state actors have enacted state-led agendas through (transnational) urban policy mobilization, foregrounding the transnational dimensions of the Singapore model and the SSTEC. Second, it attends to the historical politics of anxiety via Singapore's post-independence origins and its positionality as a small Chinese-dominated country in the "Malay world" of Southeast Asia, showing how the Singapore model became deployed by the Singaporean state as a tool of diplomacy with the Chinese state to negotiate regional power dynamics. Third, it follows the Singapore model to the symbolic terrain of the SSTEC, exploring the practitioner politics of anxiety from the perspective of Singaporean state officials which manifested in the form of intergovernmental tensions that were interpreted, crucially, as symbolic of changing power dynamics in Sino-Singaporean bilateral relations. The fourth section concludes the paper.

Governing policy failure

Presently there is a small but growing body of work in critical geographical scholarship that addresses failure, most frequently vis-à-vis policy and globalization research. While no standard definition of "failure" among critical geographers currently exists, most would agree that failure occurs when the objectives of key stakeholders are not met. Relatedly, failure is understood to emerge when policies, programs, or projects do not materialize as planned (e.g., Chang, 2017). A more expansive reading views these occurrences as symptomatic of deeper regulatory shifts in policy regimes (e.g., Brenner et al., 2010; Wells, 2014); a generalized disruption of (globalizing) flows (e.g., Perrons & Posocco, 2009); or, quite simply, the unwanted "other" of success. Generally, the literature lacks a systematic definition of what exactly "policy failure" is and how such instances of failure can be recognized across different geographical contexts. In particular, there is a troubling slippage, materially and discursively, between the variegated forms of failure that manifest during policymaking. Does "policy failure" mean the same thing as when a policy does not "proceed as expected" (Lovell, 2019, p. 314), compared to when an entire infrastructural project, typically composed of a range of policies and programs, "fail[s] to materialize" (Chang, 2017, p. 1725)? Such types of policy-related failures are qualitatively distinctive in terms of form, process, and scale. Explaining their coexistence and contradictions requires greater precision in our conceptualizations. By examining instances of "failure" from the perspectives of Singaporean state officials, this paper highlights its historicized, socially constructed nature, suggesting that one way forward is to pay attention to how practitioners define, negotiate, and contest "failure."

In that absolute failure remains more mythical than actual, megaprojects rarely fail *entirely* because the variety of policies and ideational components involved are usually mobilized elsewhere, eventually, in spite of their authors' intentions. Lauermann (2016) examines failed Olympic bids in the US, arguing that despite the failure of its bid, Boston continued to operate as an institutional intermediary for re-enrolling American policy-makers into wider Olympic networks. Chang (2017) shows that the failure of the Dongtan Eco-city to physically materialize hardly impeded subsequent teams from extracting lessons for eco-city-building through study trips, though it was likely for such reasons that Dongtan became pathologized as an "anti-model," a cautionary tale best circumvented than a success story worth simulating (Kennedy, 2016). This ambiguity of pinpointing the exact limits of failure suggests that failure and success are not cleanly separated. Embracing these entanglements, scholars have conceptualized alternative vocabularies, emphasizing process over outcome. Wells (2014, p. 475) conceptualizes "policyfailing" to locate the repeated "moments in which policies are defeated, stopped, or stalled … the making of a policy may fail temporarily, repeatedly, or permanently. What demands attention [are] … the actual practices and conditional forces that create these moments of policyfailing." To "attend to practices that fall between (and beyond) success and failure," Colven (2020) illustrates these interconnections through Jakarta's Great Garuda Seawall. Similarly, this paper conceives of "anxiety" as the framework through which ideals and instances of "success" and "failure" are relationally co-constituted and co-produced.

It is not unusual, however, to find that even if categories of success and failure are entangled in both practice and theory, these categories continue to remain distinct and politicized by officials, surfacing tensions between both perspectives. State leaders are famously reluctant to acknowledge failure, either publicly or privately, as this would require a radical rethinking of core governance values (May, 1992). To reaffirm dominant ideologies and to preserve organizational coherence, states are more inclined to "airbrush out" policy or political failures (Jones & Ward, 2002, p. 487). This heightened awareness of, and aversion to, failure assumes greater significance in cases where large-scale infrastructure becomes a "spectacular" nation-building project, physically and politically prominent, heavily symbolic of state agendas and aspirations of the future that are constitutive of the "ongoing art of being global" (Ong, 2011, p. 3), especially significant in postcolonial cities of the Global South where state-corporate alliances rework land and capital in transformative ways (Shatkin, 2019). Large megaprojects can also "introduc[e] new geopolitical dynamic[s] among countries in the region," as Moser (2018, p. 936) shows in the case of Forest City, Malaysia, where Chinese investment reworked existing regional hierarchies.

To unpack the Singaporean state's (geo)political agendas in exporting the Singapore model abroad, as well as the heightened political stakes of the SSTEC, this paper draws on research that brings a more political geographic lens to transnational urban policy mobilization, critically examining state agendas and the agency of state actors to scale up their ambitions to "shap[e] such spatial transformations to state interests" (Shatkin, 2019, p. 69). On the use of "geopolitical" as a spatial scale of analysis, I follow Sioh (2010) in acknowledging that "relationally, geopolitical intervention and negotiation take place at different levels, not just between citizens and the state but between states." This paper focuses on interstate relationships, in this case Sino-Singaporean bilateral relations from the perspectives of Singaporean officials, which are grounded in "a postcolonial government's anxiety about its legitimacy" (Sioh, 2010, p. 469). Extant work that explores the role and implications of nation-state actors in driving and shaping transnational urban policy mobilization, specifically "world-aspiring projects" (Ong, 2011, p. 4), has largely been undertaken in historical-geographical settings beyond the Global North. Bunnell (2003) analyzes how Malaysian state officials sought to advance non-Western visions of modernity through the Multimedia Super Corridor. Koch (2013) foregrounds cities as privileged sites of nation-building that "synecdochically represent the country as a whole," examining situated Kazakhstani geopolitical imaginations amongst state officials and citizens, concluding that discussions of contemporary urbanization must consider "broader (geo)political context[s] in which the discourse – encompassing material and rhetorical intervention – unfolds." Other relevant political geographic work that foregrounds nation-state actors includes the emergent research on "urban developmentalism" (Doucette & Park, 2018), which situates East Asian urbanization within wider geopolitical-economic contexts. They investigate how the ambitions of (geo)political alliances govern urban space through "spaces of exception" and "transnational networks of expertise." Collectively, these strands of work emphasize the agency of nation-state officials in transnational urban development, reinforcing the enduring significance of (geo)political relations and national politico-economic contexts.

This provides a starting point for understanding how the Singapore model is mobilized transnationally through a politics of anxiety. Through the eyes of Singaporean state officials, it illustrates the amplified political stakes that are attached to instances and outcomes of "success" and "failure." These are mutually reinforcing, driving each other through a *historical politics of anxiety*, contextualized within Singapore's history of (geo)political turbulence in Southeast Asia. Hence the Singapore model became a diplomatic tool to strengthen bilateral relations with "big powers" like post-1978 China, leading to the emergence of Sino-Singaporean megaprojects. This can be read through a South-South frame of transnational policy mobilization, an "inter-Asian horizon of metropolitan and global aspirations" (Ong, 2011, p. 5). But efforts to export the Singapore model to urban China reveal the *practitioner politics of anxiety* that manifest through the perspectives of Singaporean officials, spatialized through the SSTEC's territorial politics. These anxieties are symbolic of the changing relations between the Singapore model and its Chinese audience, redolent of a reversal of power in Sino-Singaporean bilateral relations.

The historical politics of anxiety of the Singapore "model"

> There's always a certain anxiety that our geographic, economic, and political positions are vulnerable. This anxiety is also a galvanizing force, in some ways an obsession. Our success is the result of anxiety, and the anxiety is never fully assuaged by success. Perhaps most city-states feel that way. It keeps people on the ball.
>
> – George Yeo, Singapore's former Minister of Foreign Affairs (Yeo, 1997, p. 30)

Yeo's quote aptly captures the heightened political sensitivity attached to labels such as "success" and "failure" and, more intriguingly, how the two are mutually constitutive through a politics of anxiety. Anxiety, both "galvanizing force" and "obsession," is rooted in state meta-narratives of "vulnerability" and "relevance", productive of a form of success whose momentum is (contradictorily) spurred by fears of failure. Chua (2017, p. 2) frames this positionality as a "generalized anxiety about the long-term viability of the social, economic and political foundation of the island-nation [that] has been transformed into a set of ideological justifications for and instrumental practices of tight social and political control, which taken together constitutes the authoritarianism of the regime." Anxiety characterizes an everyday mode of existence, for the Singaporean state and society, through which success and failure are doubly bound in an awkward yet generative relationship.

As a developmental narrative predicated on the substantiveness of economic achievements and absolute planning control (Shatkin, 2014), the Singapore "model" belies such sentiments of apprehension and fallibility. Over the past five decades, Singapore's varied accomplishments (e.g., rapid economic growth, public housing provision, environmental sustainability) have been considered unprecedented by conventional measures, and are regularly celebrated in international rankings. Visitors to the city-state often depart with the impression that it is a clean, safe, and well-regulated city where things, by and large, simply "work." These contribute to the sense that Singapore is a model worth emulating, aspirational notions that have struck a chord with a wide swathe of developing countries. Paul Kigame, Rwanda's president, wants Rwanda to be the "Singapore of Africa" (The Economist, 2012). The Indian government is developing a "Little Singapore" along the Delhi-Mumbai corridor (Harris, 2012). Shatkin (2014) suggests that the Singapore model is so popular in the Global South, its primary audience, precisely because it offers an alternative ideological formula of growth and governance that deliberately disavows Western-style liberal democracy. To that end, the model is "a valorised object/place ... as well as an orientation to the world" in the form of a "self-orientalized Asian success story" (Pow, 2014, p. 288).

Capitalizing on its position in global networks of urban knowledge and the eager appetites of audiences (Peck and Theodore, 2010) , the Singaporean state has risen to the occasion. If models travel in disaggregated, disembedded forms (Peck, 2011), then segments of the Singapore model have been repackaged by state agencies and state-linked consultancies for export, as discrete parts of a wider success story. State-linked institutions frame their expertise through locally-rooted, homegrown narratives of success, manifesting through calculated appeals to the "Singapore heritage of cost-effective and efficient city planning" (Surbana, 2014, p. 66), although these exports abroad inevitably generate conflicting outcomes (Bok, 2015). Still, as Chua (2011) remarks, the Singapore state remains confident enough of its methods to the extent that various institutions have been established to disseminate this brand of success. The Singapore Civil Service College was founded in 2003 to "share Singapore's experience in public reforms and good governance" (CSC, 2017), while the Lee Kuan Yew School of Public Policy emerged in 2004 to "educate and train the next generation of Asian policy-makers and leaders" (LKYSPP, 2018). Across these benign refrains of "sharing" and "educating," policy success is articulated through the commoditized and circulatory elements of the Singapore model and via the confidence of its architects, embodied and emboldened, to export their systems of governance elsewhere. One interviewee, a former diplomat, framed the Singapore model as an opportunity "to use what we've done well to win friends overseas, [since] we've become an example, a showcase … to create relationships between countries and people" (author's interview, 2014). Such are the bilateral stakes of transnational (urban) policy mobilization, wherein the Singapore model is leveraged as a diplomatic tool to build geopolitical relations.

In reality, this model is the emblematic by-product of a general ethos of success dispersed throughout the nation. "'Success has entered the process of 'self-scripting' of Singapore as a nation and of individual Singaporeans … Iconic achievements all add up, finally, to a sense of arrival at a First World economy … 'Success' as a source of pride has become part of the technologies of the Singaporean self and a constitutive element of the Singapore identity" (Chua, 2011, p. 31–32). Similar to Koch's (2013) study of how Turkic state-led developmental agendas of "modernity" are performed through the popular geographic imaginaries of elites and non-elites alike, ordinary Singaporeans take pride in – and are at pains to defend – the city-state's presumptive identity of success. This popular geographic imaginary of nationalism is not merely imposed from the top-down, but is actively reworked and reproduced from the bottom-up. The Singaporean civil service, comprising both elites and non-elites, bridges this. Yet, such an exaggerated, pronounced perception of success as integral to national identity is, correspondingly, plagued by an amplified awareness of failure, to the point where the prospect of failure becomes indistinguishable from the burden of success. For both citizen and state, anxiety is doubly leveraged, "motivat[ing] individuals to compete fiercely to maintain – better still to extend – their success [while] keep[ing] the government constantly in search of the next niche of development … [as] it needs to extend success for political legitimacy" (Chua, 2011, p. 32).

Framing anxiety as a "galvanizing" socio-political force emphasizes how it is "collective and distributed, rather than solely a property or experience of the individual" (Jackson et al., 2012, p. 24), embodied existentially. For East Asian developmental states, economic success was historically rooted in race-based discourses of "Asian values," or the cultural performance of developmental models that is entangled with postcolonial identity crises (Sioh, 2010). More than a latent sense of unease that afflicts individual state officials, fear of politico-economic failure is something of a dogma for the national ruling party. Ever-conscious of shifting global terrains underfoot, politicians from the long-ruling People's Action Party (PAP) have cautioned: "Should Singapore be overtaken and be made irrelevant, our influence and international standing will go down" (The Straits Times, 1997). Regularly disseminated through state-owned media, fears of failure are expressed as exhortations for Singaporeans and Singapore to remain "relevant," the PAP's doctrine. One interviewee bluntly said: "If you are relevant, you have a reason to exist. If you don't know anything of use to anyone, you are forever cut out of the market" (author's interview, 2015). Modeling, as a success-oriented technology of governance, is driven by the enterprise of (developmental) state capitalism: it is entrenched in present and future ideologies of national socio-economic growth, directed at cultivating desirable citizen-subjects, and enacted through regulatory technologies (Hoffman, 2011). For the PAP, prospects of politico-economic decline convey a palpable sense of anxiety, wherein success and failure are articulated through the existential tones of relevance and irrelevance to the world.

This political ideology of vulnerability that permeates state agendas is historically rooted in Singapore's expulsion from Malaysia in 1965, and the antagonism it received from Indonesia in the aftermath of *Konfrontasi* during the 1960s, in the wake of British decolonization. The idea of a sovereign Singapore then was considered a "political, economic, and geographical absurdity," primarily because the continued existence of a small island city-state that lacked a hinterland, and was populated predominantly by overseas ethnic Chinese amidst the "Malay world" of Southeast Asia, namely its neighbors Malaysia and Indonesia, was deemed an impossibility (Acharya, 2008). Official representations of that "founding moment" continue to paint it as a traumatic political experience, serving to define a national predicament and justify a national watchword of "vulnerability" that remains "continually confronted by prospects of political extinction" (Leifer, 2000, p. 1). The state has never taken Singapore's sovereignty for granted. The expulsion produced a deep security complex that continues to shape the conduct and culture of its foreign policy to negotiate vulnerability as a small city-state. An interviewee, another diplomat, stressed: "Singapore is a city-state – you cannot say that if the city fails, you have the hinterland. If Singapore fails, there's *nothing* left" (author's interview, 2014). Such is the utter existentialism of (the prospect of) failure, politically amplified and inseparable from the sovereignty of a small state.

Within the wider Southeast Asian and Pacific Asian region, fears of conflict with its closest neighbors, Malaysia and Indonesia, remain longstanding preoccupations of the Singapore state. The sense of national confidence engendered by success also expresses itself as a "quality of hubris" that arouses "admiration mixed with envy and resentment within [Singapore's] regional locale, where its success has been represented at having come at the expense of its close neighbors" (Leifer, 2000, p. 41). This explains why the Singapore model is consistently viewed with suspicion by Malaysia and Indonesia, who remain distrustful of Singapore's "neocolonial" economic aspirations (Bunnell et al., 2006). It underscores the uneven reception of models, together with the compounded awareness of political vulnerability and anxiety that, for this small city-state, is interlocked with success. The (geo)politics of anxiety are both historically predicated upon, and contradictorily productive of, the intertwining of success and failure.

The Singapore state is constantly searching for new ways to ameliorate political anxieties. Over the last three decades, the state's export of the Singapore model has been framed as a diplomatic opportunity to build and strengthen bilateral relationships with "big powers" in the region, most notably China. An interviewee brought up the Singapore model, saying: "We want to engage China, using what we've done well as a way to make good friends" (author's interview, 2014). This was a common refrain over the interviews. Sino-Singapore geopolitical relations are constituted through a stream of policy exports within customized institutional architectures that aim to promote substantive exchanges of expertise and interpersonal interactions between (ideally) high-ranking state officials. Beginning as a tentative enterprise, this bilateral relationship grew from the Singaporean state's interest in maintaining a regional balance of power that would deny undue dominance to any one country (Singh, 1988). Sino-Singapore relations were tense in the 1950s and 60s owing to China's support of Communist insurgency movements in Southeast Asia, which threatened to overthrow postcolonial governments in the region. As a newly independent nation, Singapore's stance was to ease the concerns of its neighbors, who were wary of Chinese intentions in Southeast Asia (Acharya, 2008). A turning point came in the form of Richard Nixon's visit to Beijing in 1972, which impelled Deng Xiaoping, the-then leader of the Chinese Communist Party (CCP) to visit Southeast Asia in 1978. Deng was apparently impressed by the social and economic progress Singapore had attained since independence, marking the start of positive portrayals of Singapore in Chinese state media as "a garden city worth studying for its greenery, public housing and tourism" (Lee & Yew, 2000, p. 65), for Deng had seen for himself that it was possible to combine one-party governance with economic growth. In 1992, Deng embarked on his "southern tour" around China, during which he explicitly singled out Singapore as a model: "Singapore's social order is rather good. Its leaders exercise strict management. We should learn from their experience, and do a better job" (Kristof, 1992). The CCP duly dispatched a senior delegation to Singapore, returning with proclamations of Singapore as a model to be emulated, thereby instantiating a deeper engagement between the two political regimes. This would set in motion tens of thousands of "learning" visits by Chinese delegates to Singapore over the next three decades, highlighting not just the presence of organized and institutionalized policy exchange, but also a "wider cultural diplomacy ... strikingly different from ... policy tourism today" (Cook et al., 2014, p. 818).

For its own part, modeling has been a longstanding element of Chinese rule since the 1920s to experimentally advance agendas and shape society. Bakken (2000) documents how model citizens and sites have been upheld as exemplars to emulate, signposts for where Chinese society should be heading. More internationally, the CCP has turned from Soviet-inspired practices of industrial economic development to contemporary models based on cities such Kitakyushu and Singapore (Hoffman, 2011). The choice of which places to anoint as "models" remains politically driven. As a formerly socialist country undergoing capitalist reform, China was reluctant to uphold an advanced capitalist nation as a developmental model. And neither Taiwan nor Hong Kong, which continue to be viewed as integral elements of Chinese sovereignty, was considered suitable (Wong & Goldblum, 2000). As a post-colony and a small, non-threatening nation, Singapore was an ideologically suitable option. One interviewee matter-of-factly remarked, "Nobody feels threatened by Singapore. If this were Indonesia or Malaysia, China would say, what are you trying to do? But we are just so *small*" (author's interview, 2014). For Beijing, therefore, Singapore represented the capitalist counterpart of the Communist dream (Cartier, 1995), a developmental narrative *par excellence* orchestrated by an authoritarian state with a history of single-party rule. The choice of Singapore was therefore a politically mediated one, congruent with dominant Chinese regime goals.

Amongst the range of Sino-Singaporean bilateral exchanges over the last three decades, the export of the Singapore model to urban China, through inter-governmental infrastructural megaprojects such as the SSSIP and the SSTEC, has been lauded as the pinnacle of Sino-Singaporean bilateral relations. State officials use the terminology of "software transfer" to characterize the export of Singapore's "accumulated and proven methods of development and administration ... [based on] its successful experiences" to Chinese government, such that Chinese officials might "understand the Singapore way ... and decide how best to adapt Singapore's practices for local circumstances" (Pereira, 2002, p. 129). Similar to Easterling's (2014, p. 26) software metaphor which depicts how "the dominant formula that generates Shenzhens and Dubais" is a particular way of "making urban space" in the likeness of someplace else, G-G projects operate on the assumption that the Singapore model can more or less be easily transplanted into foreign settings, allowing Chinese policymaking to "capture some aspect, style, or essence of that original" (Ong, 2011, p. 15). This "formula," or the performance of a rational, methodical transfer of policy expertise, is enacted through training programs and learning tours, during which Chinese state officials are trained by their Singaporean counterparts in different aspects of development.

G-G projects are governed through a distinctive institutional architecture that prioritizes heavy state (central and municipal) involvement and political resources in order to ensure that the project comes to fruition. Collaboration occurs at two main levels. At the strategic level, a Joint Steering Committee and a Joint Working Committee meet every three months to chart project overviews and implementation. At the operational level, Chinese and Singaporean policymakers have formed administrative committees to supervise key domains of urban development (e.g., urban planning, environmental protection). Such "organizational cultures" (Schäefer, 2017), or state structures, shape policy mobilization path-dependently by generating ideological and institutional legacies across political generations. Interviewees emphasized the unusually high levels of state support that were channeled toward the SSTEC. An economic planner noted: "With G-G projects, you know there's the commitment to implementation, unlike private sector projects" (author's interview, 2014). This political "commitment" suggests that outcomes of the SSTEC symbolize the health of Sino-Singaporean bilateral relations. With these "exception[al]" (Caprotti, 2015, p. 36) state resources, the assumption for Singaporean state officials was that if the Singapore model could be successfully exported anywhere, it would be under such circumstances. The SSTEC is interesting, therefore, in that it carries an almost *presumptive* air of success, within which fears of failure are cloaked. Singaporean officials labor under the belief that *if* the Singapore model could be successfully exported anywhere, it would be under such interstate-enabled circumstances in China. They ask, almost rhetorically: if not here, then where?

The practitioner politics of anxiety in the Sino-Singapore Tianjin eco-city

"There's always a risk. But there is the impression that this is an inter-governmental project, it's not allowed to fail, so the risk, compared with going to other places in China, might be lower," a Singaporean official explained to me during discussions of how firms might be persuaded to invest in the SSTEC. When pressed to elaborate on what "not allowed to fail" meant, the official emphasized the SSTEC's special status. How many top Singaporean civil servants, she asked, could be channeled into a project on foreign territory? Doing so would show the Chinese leadership how much the Singaporean state valued the project. Others commented that the progress of the SSTEC would reflect on Sino-Singaporean bilateral relations. On the surface of it, and largely related to the fact that it has undergone physical materialization amidst a sea of unfinished projects, the SSTEC impresses upon onlookers near and far ideals of "success." Furthermore, the SSTEC possesses green building credentials that are yoked to international standards, such as LEED, and boasts its own Key Performance Indicator framework (SSTEC, 2019), comprising benchmarks and deliverables. Such metrics project future progress and invite comparative emulation. Singaporean officials regularly feature the SSTEC in conferences and learning tours for developing countries, as evidence for how (and where) the Singapore model has "worked," constituting the projective "knowledge effects" (Merry, 2011) that radiate the banal essence of success through the performative power of repetition. Such approaches have proved salient in Chinese urbanization, where the "design of eco-cities in China has been distinctively characterized by an abundance of images as a form of communication of ideas" (Morera, 2017, p. 190).

The major source of "practitioner anxiety" stems from changing Sino-Singaporean power dynamics between a developmental model and an audience that was systematically courted for decades. Doucette (2020, p. 6) contextualizes "practitioner anxiety" in how "the meaning of Korean development occupies a zone of awkward engagement", complicating the seeming straightforwardness of knowledge-sharing by examining how the "anxieties that shape specific efforts to construct and circulate development experience narratives afflict those tasked with sharing them ... show[ing] how new connections and flows in development cooperation and knowledge sharing also create anxious and awkward engagements." For Doucette (2020, p. 3), anxiety assumes the "mundane form of uncertainty among professionals ... [and] everyday subjective feelings and reflections." Every practitioner arguably faces some anxiety during projects when things inevitably go awry. Singaporean state officials experience anxiety via banal feelings of uncertainty, resignation, and thinly-veiled antagonism toward the actions of their Chinese counterparts. Situated in the historical politics of anxiety experienced by a small city-state that is dealing with a "big power," such interstate interactions take on an existential bent. In other words, state anxiety and practitioner anxiety become nearly indistinguishable in the context of a (geo)political backdrop against which Singaporean officials experience feel a palpable sense of expiration regarding their "relevance" for Chinese officials, culminating in the sense that the Singapore model is now appreciated more for its performance of success rather than any functional prospects of genuine policy exchange and learning. Examining "anxiety in practice" (Doucette, 2020) from the perspective of Singaporean officials shows how geopolitical scripts of state "vulnerability" and "relevance" are reproduced and grappled with on the symbolic terrain of the SSTEC.

As the predecessor of the SSTEC, the fate of the SSIP warrants some mention in order to contextualize the practitioner anxiety faced by Singaporean officials during the SSTEC project, as they negotiated the SSSIP's legacy of an embarrassing "failure." As the first Sino-Singaporean G-G project, the SSSIP was established in 1994, boldly hailed as a pioneering enterprise in software transfer to export the Singaporean state's expertise in urban-industrial development to Chinese government (Pereira, 2002). By the early 1990s, though, there were hundreds of industrial estates in Chinese city-regions developed by municipal governments, an early hint of the scalar disjunctures between Singaporean and Chinese political economies. Nonetheless, the Singaporean state remained confident that the SSSIP possessed distinctive competitive advantages in the form of the "Singapore brand name," an offshoot of the Singapore model that indicated "policy credibility" (Huff, 1995) to assure investors new to post-1978 China that the park was of reliable, international standards. Distilled into various forms of "reputational capital" (Phelps, 2007), the Singapore model was viewed by Chinese state officials as crucial for territorially embedding investments in Chinese city-regions. The SSSIP initially took off, but started suffering financially three years later owing to its inability to compete with a rival industrial estate: the Suzhou New District. The Suzhou Municipal Government originally offered this site for the location of the SSSIP but the Singaporean state declined, preferring to develop its own project from scratch. Ironically, the rival district's success stemmed from the Singapore state's effectiveness in exporting the Singapore model to Suzhou. Chinese officials from the District were able to "observe, absorb and reproduce the [SSSIP's] 'international standard' practice ... [and were] quick to learn how to market the estate internationally" (Pereira, 2002, p. 132). The Singaporean state officially left the SSSIP in 2001, selling 30% of its shares to the Chinese consortium to give the latter majority ownership. For Singaporean officials, the key indicator of failure was the dramatic exit of the Singapore state from what had been trumpeted as a groundbreaking inter-governmental project. The "Suzhou fiasco" (Pereira, 2002, p. 141), labeled as such by the public for what was perceived as a wastage of resources beyond Singapore's domestic confines, triggered defensive responses from state officials, in turn revealing the difficulties of exporting the Singapore model elsewhere and the domestic dissent generated by "failure" abroad.

From the perspectives of Singaporean state officials, this exit was framed as a key rupture of the SSSIP, highlighting two instances of failure during this inaugural effort to export the Singapore model to China: the inability to overcome interlocal competition; and the incapacity to comprehend central-local Chinese state politics. The first concerned a short-termist, entrepreneurial outlook in Chinese municipal governance that continues to be tied to the generation of fiscal revenue and political legitimacy, termed "GDP-ism" (Wu, 2016). Following expanded devolution and market entrepreneurialism, the performance of municipal officials has been evaluated through their ability to meet growth targets set by the central state, hence the emergence of competing developments and territorial politics that indicate uneven local state control over land (Chien & Gordon, 2008). The second arose from the unfamiliarity of Singaporean policymakers – who are accustomed to "micro-manag[ing]" (Lim Kean Fan, & Horesh, Niv, 2016, p. 1005) social, political, and economic matters at the urban-national scale (Olds & Yeung, 2004) – in negotiating the spatio-scalar complexities of Chinese administration. It has been acknowledged by the PAP and external critics that Singaporean officials overly focused on cultivating personal relationships with central state officials; this came at the expense of relations with municipal state officials, who were misjudged by Singaporean officials as incapable of acting independently of the central state. This scalar disjuncture reflected fundamental contradictions of the attempt to export a model from

a city-state to a multi-level bureaucracy. For Singaporean state officials, the most pertinent question surrounding the SSTEC was whether "the Suzhou experience would re-
emerge" (Lim & Horesh, 2016, p. 1008). Would the SSTEC, as the second concerted
inter-governmental effort to export the Singapore model abroad, be tainted by associations with failure? These compounded fears, coupled with already-existing (geo)political
insecurities embodied by state officials, took the form of practitioner-based anxieties that
manifested during the SSTEC project. If the historical politics of anxiety previously
discussed were explicitly existential, then the practitioner anxiety that surfaces here can
be interpreted as (geo)political existentialism in practice.

The Singaporean officials with whom I interacted displayed, almost consistently,
a heightened sensitivity to the topic of transnational policy mobilization, especially
vis-à-vis sovereignty, revealing attitudes of precarity when dealing with the world
from the positionality of a small state. A senior Singaporean official who had led the
SSTEC's green building policy was triggered by my mention of "policy transfer,"
remaining adamant that policy could not be "transferred" for reasons of national
sovereignty:

> Since this is a joint venture, the Chinese will not want to adopt the Green Mark [Singapore's
> official green building policy]. As a sovereign nation, we don't want other countries impos-
> ing policies on us ... it's national pride. (author's interview, 2014)

The vernacular of "transfer," amplified through the precarious dealings between small
and big states, was framed as a potential threat to bilateral relations. Other Singaporean
officials seemed wary of (verbally) overstepping political boundaries surrounding ostensibly benign notions of policy sharing and learning. Many demurred to position
Singapore as a "model" or to use terms like "educate" in the context of policy learning.
Curiously, this reticence starkly contrasts more confident, emphatic declarations at
closed-door workshops on the Singapore model and its global viability. I have witnessed
the inflationary posturing of Singaporean officials during encounters with smaller cities
in developing Southeast Asian countries, such as the Philippines and Vietnam, when it
comes to "selling" the Singapore model and summarily peddling the portfolios of
Singaporean planning consultancies. These divergences constitute the mobilities of
urban models, or a geography of scripted superiority. The comparatively deferential
attitude toward the Chinese state makes sense when situated within small-big state
relations, especially considering how the Chinese audience has been systematically
cultivated for decades in a fashion unusual for any other country in Singapore's foreign
policy. Herein lie the power dynamics in policy mobilization that are inextricable from
positioning one party as a "teacher" and another as a "learner." If anxiety in practice can
be interpreted as the uneven manifestation of a latent quality of defensiveness, then the
context in which anxiety rears its head is a politically significant question that foregrounds power in social relations, complicating literalist notions of policy "transfer" to
account for nationalist concerns.

More prominently, practitioner anxiety results from how Singaporean officials have
perceived the changing attitudes of Chinese officials toward the Singapore model from
the SSSIP to the SSTEC. Singaporean officials reinforced their fears over these changing
dynamics, which signified a reversal in dependence. A junior official reflected on how "we

still have quite a privileged position with China, although that's diminishing." When asked to elaborate on this "diminishing" position, she replied:

> As China opens up, it has more direct channels with other parts of the world; it doesn't need to keep depending on Singapore. During Suzhou we were the bridge … as China rises, it needs us less and less … That's why we are more dependent on them. (author's interview, 2014)

A retired senior official who had been involved in both projects compared the shifting power dynamics:

> At this point of China's development they are no longer like they were during the Suzhou days …. they were very *blur*, no money, no track record, no expertise – which is why, in Suzhou, Singapore took complete charge … basically they listened to us. Tianjin came 15 years later. Now the Chinese have money, quite well-established. They are full of confidence, everyone is going into China. The Chinese jokingly tell us, why do we need you? You are so small, we are so big. (author's interview, 2014)

These quotes are from senior and junior officials, indicating the reproduction of practitioner and state anxiety, along with a tacit acceptance of state (geo)political scripts of "vulnerability" and "relevance" to frame Singapore's place in the world. Where Singaporean officials once saw themselves performing the integral role of a "bridge" between an emergent China and the rest of the world – a role that must have seemed phenomenal to a small city-state – they now have to contend with a growing realization that the tables have turned. The asymmetry in dependency that was apparent during the SSSIP, and which has progressively shrunk, is in fact the contradictory power relations of a model laid bare. One of the ultimate contradictions of a model, at least in practice, is that if a model works to the genuine learning advantages of its audience, its utility in turn declines. After all, successful policy transfer is fundamentally contingent upon, and reproduced through, a perceived asymmetry between the parties concerned, notwithstanding changing ideological parameters of "success" itself (e.g., Zhang, 2012 on Shenzhen's shifting aspirations from Hong Kong to Singapore). Arguably, it was only a matter of time before these global shifts came home to roost, raising the temporal question of how long the architects of a model might retain its legitimacy for audiences – hence the conundrum of expiration, or when model co-production dissolves. There remain attempts, nonetheless, to assuage anxiety by continuing to engage Chinese audiences through highlighting new developments in Singapore, since "we have to pursue them and show our relevance … otherwise we will be left behind" (author's interview, 2014).

Practitioner anxieties manifested spatially, and perhaps most dramatically, through land disputes and shifting territorial boundaries, resulting in crucial disruptions and delays. Publicly, the SSTEC is framed as a bilateral collaboration between the Singaporean and Chinese central governments; on the ground, its fate is determined by the subnational governments of the Tianjin Municipality and the regional Tianjin-Binhai New Area (TBNA) that oversee its everyday operations. The SSTEC was originally planned as a stand-alone 3 km2 site – with a dedicated "green" transportation system – but conflicting interstate priorities and attempts by different levels of Chinese government to integrate the SSTEC into wider regional configurations have complicated matters. During the initial negotiations, it was agreed that the Singaporean state's efforts would be limited to the 3 km2 site. Subsequently, to capitalize on collective Sino-Singaporean efforts and resources, the TBNA government unilaterally decided that the

SSTEC's territorial boundaries should be expanded to include a tourism district. A Singaporean official said, exasperatedly:

> We conveyed that our collaboration is still within the 3km2 ... we shouldn't even be debating. But this is the truth about working with the Chinese. They will change, so we need to stand firm ... The tourism district did their own masterplan and now they say it's under the eco-city. But it's not. We need to stay focused. (author's interview, 2014)

Another complained about this instability of Chinese policymaking, which was interpreted as an unwelcome yet characteristically entrepreneurial act of "piggybacking" onto interstate efforts that had garnered traction: "They were already planning the tourism district alongside the eco-city, but they only told us later ... Once the Chinese see you taking off and doing well, they will redraw the masterplan to help adjoining cities" (author's interview, 2014). Several officials noted that the biggest challenge was the establishment of extensive transportation linkages within the SSTEC itself, and between the SSTEC and Tianjin Municipality. They expressed frustration over the delays experienced in convincing Chinese officials to commit to building these forms of infrastructure, and bemoaned the loss of eco-friendly transportation which they saw as critical to the SSTEC's measure of "greenness." These disruptions in transportation planning, resultant of the Singaporean state's continuing inability to negotiate scalar disjunctures in Chinese governance, had wider repercussions. They triggered the departure of leading international developers that were needed to develop the SSTEC, demonstrating how the SSTEC's "greenness" depended on the inclusion of developers to transfer green technology across space (Bok & Coe, 2017), but also reflecting the speculativeness of new-build developments more generally (see Upadhya, 2020). A Singaporean interviewee, who was the pioneering CEO of the SSTEC, was the most disgruntled of the lot, viewing their loss as the reason why he considered the project:

> ... a super failure. Because transportation is gone, many things have fallen aside. The developers signed, thinking there'd be transport, but they've now all left. Development is stuck in the first phase. When they gave up my LRT [transportation network], it was the stupidest thing ... Even my own government was on their side! You should be fighting for the interests of Singapore! When I look back, this project helped me make up my mind to leave the civil service. (author's interview, 2015)

Behind the benign performance of mutually beneficial interstate projects, wherein South-South exchanges are popularly portrayed as "idealized, if imaginary, concept[s] of horizontal, more equal, win-win interaction[s]" (Horner, 2016, p. 402), old tensions rear their heads. Here, changing power dynamics in Sino-Singaporean bilateral relations, and the corresponding reversal in relations of dependence, emerged in direct conflicts not just between states, but within the Singaporean delegation itself regarding how much Singaporean officials were expected to acquiesce during bilateral negotiations. While most interviewees chose to ascribe feelings of resignation to the realities of how a small state should behave when working with a "big power," a few went against the grain in order to secure what they felt was "in Singapore's interest." Other than surfacing an unacknowledged streak of nationalism during transnational urban policy mobilization, this also illustrates the range of attitudes within the "institutional ensembles" that are state delegations (Jessop, 2016). This ex-CEO also commented on land disputes during the early stages of planning:

> I chanced upon a masterplan on my Chinese consortium Chairman's table. He cut it out, like a warlord, for big Chinese developers. He didn't share it with me, but I could see all the Chinese developers and their friends claiming plots of land. I was furious; I felt trapped. If I didn't work harder, they would sell all the land. And when you fail to bring in investors, the Chinese will say, don't worry, I help you [laughs]. (author's interview, 2015)

Ostensibly, the main selling point of the SSTEC was the Sino-Singaporean central state backing and technocratic expertise that was intended to be collaboratively channeled into every stage of the project. For some Singaporean officials, however, it seemed that their Chinese counterparts were, in a sense, already planning for failure by unilaterally allocating prime land in ways they considered more suitable, revealing the hollowness of bilateral scripts of mutual benefit. This preemptive position regarding the seeming utility of Singaporean policy expertise was echoed by another Singaporean official who had led the SSTEC's masterplanning:

> For the first meeting, we met our Chinese partners, but we got a huge shock. They *kay kiang* [Singaporean colloquium to describe rashness] and already came up with their own mas-terplan! It was done by the Tianjin Planning Academy – they also brought in the national planning academy. (author's interview, 2014)

These instances cast ambiguity upon the specific role of the Singapore model for an intergovernmental project whose credibility was, and continues to be, marketed on this very basis of Singapore-branded urban expertise. Apart from the acknowledged perfor-mativity of masterplanning artifices and architects, they also raise questions surrounding claims of (substantive) policy exchange and learning that are trumpeted during such projects. The difficulty of making qualitative distinctions between deliberate learning and more superficial forms of inter-referencing implies that "learning" remains a politically and temporally constrained and filtered process (Peck, 2011). Such barriers are suggestive of deeper ideological underpinnings that transcend individual policies or projects (May, 1992). Here, they must be situated in changing Sino-Singaporean bilateral relations, predicated on the knowledge that Chinese audiences no longer view the Singapore model as favorably as before. An official drily commented: "They still need the Singapore 'brand,' but they don't want you to cause trouble. They can do everything on their own" (author's interview, 2014). Compared to the SSSIP, when Singapore "took complete charge," the SSTEC relies on a more hollowed-out version of the Singapore model as a token of deference to certain international norms. Rather than any real interest in Singapore, Chinese engagements with the Singapore model are expressed through the motivation of advancing inward-looking agendas through profuse inter-referencing (Lim & Horesh, 2016). Sioh (2010, p. 582) references a somewhat similar state of emptiness regarding the economic success of Asian developmental states in the wake of the 1997 Asian Financial Crisis: such models were "'hollowed out' by the not-said – the anxieties and assumptions underpinning and influencing policy decisions," reflective of postcolonial state policymaking. Ultimately, nearly three decades of Sino-Singaporean policy interchanges have culminated in the reverse-engineering of a developmental model, whose architects can be kept backstage until they are wheeled out at international conferences, bilateral photo-ops, and policy tours.

Conclusion

"The window is closing" is a metaphor that was constantly invoked by Singaporean officials, encapsulating the shrinking asymmetry in Sino-Singaporean relations of knowledge and power. Grimly aware of this fast-eroding scope of "relevance," the Singaporean state is shifting its sights to new pastures of knowledge-sharing to bolster its bilateral alliances with other big powers:

> The Chief Minister of Andra Pradesh visited recently. We ... *entertain* them, so to speak. After China, window closing, right? India is five times worse but you still have to do it. You cannot align yourself with one big power and completely neglect another. India will say, "You built a few cities in China, why are you not doing one for me?" (author's interview, 2014, original emphasis)

These shifting (geo)politics of knowledge mobility reflect how a small city-state, once again, reorients itself in relation to another "big power," leveraging its expertise in urban development to forge bilateral alliances anew. Using the case of the Singapore "model" and its export to urban China via the bilateral vectors of the SSTEC, this paper has argued that categories of "success" and "failure" are relationally co-produced through a politics of anxiety. Situated in Singapore's positionality as a small island-city-state in Southeast Asia, "anxiety" captures its state of being, wherein the burden of success – almost an affliction – and the fear of failure become nearly indistinguishable. As explored above, these dynamics are expressed through the *historical politics of anxiety*, where Singapore's particular form of success and bilateral engagements are shown to be driven by anxiety over its positionality in Southeast and Pacific Asia, and through the *practitioner politics of anxiety*, where anxiety manifests through the practices of Singaporean officials in the development of the SSTEC. The Singapore model, a signifier of "success" in the Global South, is framed as a (geo)political mode of engagement between the Singaporean and Chinese states, reinforced by exports abroad that work to reaffirm (or not) the broader legacy of the model. The instances of "failure" experienced by Singaporean officials in the SSTEC capture not only the ambivalence of "success" itself but, more significantly, how what is considered as "failure" transcends individual projects and is in reality reflective of broader (geo)political circumstances, culminating in the "hollowing out" of a signifier of success.

The paper therefore advances existing scholarship in two ways. First, countering critiques of success-centrism in policy mobility research, it illustrates how success and failure are relationally bound (see Lovell, 2019; McCann & Ward, 2015). Second, in response to critiques of presentism (see Clarke, 2012; Jacobs, 2012), it delivers a historical treatment of policy mobilization by historicizing the emergence of the Singapore model and by investigating a case of South-South state projects of policy mobilization, charting the decline of the Singapore model in urban China over time.

Several overlapping conclusions for future research emerge. The interplay between occurrences of failure in the SSSIP and SSTEC projects, within the wider Sino-Singaporean policy relationship, indicates the path-dependent quality of significant failures to persist beyond projects. This reinforces the need for policy mobilities scholarship to pay attention to the structural contexts – specifically, state-level geopolitical relationships – of transnational policy mobilization. The politics of anxiety here are relatively distinct to Singapore's fairly unique positionality as a small city-state that has embarked on transnational dealings with big powers, but they call attention to how the broader positionality of states can influence nationally-inflected relations of power during transnational policy mobilization. Additionally, this case highlights the limits to (assumptions of) post-failure policy learning, which requires long-term investigation in order to critically assess the claims of actors that seek to preserve a model's legitimacy.

Acknowledgments

This paper has benefitted greatly from comments by Winston Chow, Matt Wade, Mayee Wong, and three anonymous reviewers, together with more general conversations with Emma Colven about policy mobilization. I am very grateful to John Lauermann and Cristina Temenos, the editors of this Special Issue, for their guidance and patience. I also thank Susan Moore for her editorial support. This paper has been years in the making, a product of earlier research I had undertaken during my time in the Masters program at the National University of Singapore (2013–2015). I acknowledge the funding support of the NUS Graduate Research Support Scheme. I thank Neil Coe for his supervi- sion at NUS, as well as David Sadoway for the opportunity to present these findings at a 2016 work- shop, entitled "Memes, schemes and dreams: Singapore urban futures," where the paper received some initial feedback. I dedicate this paper to the memory of Shirley Leow Sweet Mai (1935–2019).

Disclosure statement

No potential conflict of interest was reported by the author.

References

Acharya, Amitav. (2008). *Singapore's foreign policy: The search for regional order*. World Scientific.

Bakken, Børge. (2000). *The exemplary society*. Oxford University Press.

Bok, Rachel. (2015). Airports on the move? The policy mobilities of Singapore Changi airport at home and abroad. *Urban Studies*, *52*(14), 2724–2740. https://doi.org/10.1177/0042098014548011

Bok, Rachel, & Coe, Neil. (2017). Geographies of policy knowledge: The state and corporate dimensions of contemporary policy mobilities. *Cities*, *63*, 51–57. https://doi.org/10.1016/j.cities.2017.01.001

Brenner, Neil, Peck, Jamie, & Theodore, Nik. (2010). Variegated neoliberalization: Geographics, modalities, pathways. *Global Networks*, *10*(2), 182–222. https://doi.org/10.1111/j.1471-0374.2009.00277.x

Bunnell, Tim. (2003). *Malaysia, modernity and the multimedia super corridor: A critical geography of intelligent landscapes*. Routledge.

Bunnell, Tim, Muzaini, Hamzah, & Sidaway, James. (2006). Global-city frontiers: Singapore's hinterlands and the contested socio-political geographies of Bintan, Indonesia. *International Journal of Urban and Regional Research*, *30*(1), 3–22. https://doi.org/10.1111/j.1468-2427.2006.00647.x

Caprotti, Federico. (2015). *Eco-cities and the transition to low carbon economies*. Palgrave Macmillan.

Cartier, Carolyn L. (1995). Singaporean investment in China: installing the Singapore model in Sunan. Chinese Environment and Development, 6(1-2): 117–44

Chan, Heng-Chee. (1971). *Singapore: The politics of survival, 1965-1967*. Oxford University Press.

Chang, Catherine. (2017). Failure matters: Reassembling eco-urbanism in a globalizing China. *Environment & Planning A*, *49*(8), 1719–1742. https://doi.org/10.1177/0308518X16685092

Chang, Catherine, Leitner, Helga, & Sheppard, Eric. (2016). A green leap forward? Eco-state restructuring and the Tianjin-Binhai Eco-city model. *Regional Studies, 50*(6), 929–943. https://doi.org/10.1080/00343404.2015.1108519

Chien, Shiuh-Shien, & Gordon, Ian. (2008). Territorial competition in China and the west. *Regional Studies, 42*(1), 31–49. https://doi.org/10.1080/00343400701543249

Chua, Beng-Huat. (2011). Singapore as model: Planning innovations, knowledge experts. In Ananya Roy & Aihwa Ong (Eds.), *Worlding cities: Asian experiments and the art of being global* (pp. 29–52). Blackwell.

Chua, Beng-Huat. (2017). *Liberalism disavowed: Communitarianism and state capitalism in Singapore*. Cornell University Press.

Clarke, Nick. (2012). Urban policy mobility, anti-politics, and histories of the transnational municipal movement. *Progress in Human Geography, 36*(1), 25–43. https://doi.org/10.1177/0309132511407952

Colven, Emma. (2020). Thinking beyond success and failure: Dutch water expertise and friction in postcolonial Jakarta. *EPC: Politics and Space.*

Cook, Ian, Ward, Stephen, & Ward, Kevin. (2014). A springtime journey to the Soviet Union: Postwar planning and policy mobilities through the iron curtain. *International Journal of Urban and Regional Research, 38*(3), 805–822. https://doi.org/10.1111/1468-2427.12133

CSC (2017). *About us*. Retrieved from: https://www.cscollege.gov.sg/About%20Us/Pages/Default.aspx

Dent, Michael. (2001). Singapore's foreign economic policy: The pursuit of economic security. *Contemporary Southeast Asia, 23*(1), 1–23. https://doi.org/10.1355/CS23-1A

Doucette, Jamie. (2020). Anxieties of an emerging donor: The Korean development experience and the politics of international development cooperation. *EPC: Politics and Space.* online first. DOI: 10.1177/2399654420904082

Doucette, Jamie, & Park, Bae-Gyoon. (2018). Urban developmentalism in East Asia: Geopolitical economies, spaces of exception, and networks of expertise. *Critical Sociology, 44*(3), 395–403. https://doi.org/10.1177/0896920517719488

Easterling, Keller. (2014). *Extrastatecraft: The power of infrastructure space*. Verso.

Government of Singapore (2015). *National Day Rally*. Retrieved from http://www.pmo.gov.sg/national-day-rally-2015

Harris, Andrew. (2012). The metonymic urbanism of twenty-first-century Mumbai. *Urban Studies, 49*(13), 2955–2973. https://doi.org/10.1177/0042098012452458

Hoffman, Lisa. (2011). Urban modeling and contemporary technologies of city-building in China: The production of regimes of green urbanisms. In Ananya Roy & Aihwa Ong (Eds.), *Worlding cities: Asian experiments and the art of being global* (pp. 55–76). Blackwell.

Horner, Rory. (2016). A new economic geography of trade and development? Governing south-south trade, value chains and production networks. *Territory, Politics, Governance, 4*(4), 400–420. https://doi.org/10.1080/21622671.2015.1073614

Huff, W. G. (1995). What is the Singapore model of economic development?. Cambridge Journal of Economics 19, 735–759.

Jackson, Peter, Watson, Matthew, & Piper, Nicholas. (2012). Locating anxiety in the social: The cultural mediation of food fears. *European Journal of Cultural Studies, 16*(1), 24–42. https://doi.org/10.1177/1367549412457480

Jacobs, Jane. (2012). Urban geographies I: Still thinking cities relationally. *Progress in Human Geography, 36*(3), 412–422. https://doi.org/10.1177/0309132511421715

Jessop, Bob. (2016). *The state: Past, present, future*. Polity.

Jones, Martin, & Ward, Kevin. (2002). Excavating the logic of British urban policy: Neoliberalism as the "crisis of crisis-management.". *Antipode, 34*(3), 473–494. https://doi.org/10.1111/1467-8330.00251

Kennedy, Sean. (2016). Urban policy mobilities, argumentation and the case of the model city. *Urban Geography, 37*(1), 96–116. https://doi.org/10.1080/02723638.2015.1055932

Koch, Natalie. (2013). Why not a world city? Astana, Ankara, and geopolitical scripts in urban networks. *Urban Geography, 34*(1), 109–130. https://doi.org/10.1080/02723638.2013.778641

Kristof, Nicholas. (1992). China sees Singapore as a model for progress. *The New York Times*.

Lauermann, John. (2016). Temporary projects, durable outcomes: Urban development through failed Olympic bids? *Urban Studies, 53*(9), 1885–1901. https://doi.org/10.1177/0042098015585460

Lee, Kuan, & Yew. (2000). *From third world to first: The Singapore story, 1960-2000*. Times.

Leifer, Michael. (2000). *Singapore's foreign policy: Coping with vulnerability*. Routledge.

Lim, Kean Fan, & Horesh, Niv. (2016). The "Singapore fever" in China: Policy mobility and mutation. *The China Quarterly, 228*, 992–1017. https://doi.org/10.1017/S0305741016001120

LKYSPP (2018). *About us*. https://lkyspp.nus.edu.sg/explore-lkyspp

Lovell, Heather. (2019). Policy failure mobilities. *Progress in Human Geography, 43*(1), 46–63. https://doi.org/10.1177/0309132517734074

May, Peter. (1992). Policy learning and failure. *Journal of Public Policy, 12*(4), 331–354. https://doi.org/10.1017/S0143814X00005602

McCann, Eugene, & Ward, Kevin. (2015). Thinking through dualisms in urban policy mobilities. *International Journal of Urban and Regional Research, 39*(4), 828–830. https://doi.org/10.1111/1468-2427.12254

Merry, Sally. (2011). Measuring the world: Indicators, human rights, and global governance. *Current Anthropology, 52*(S3), S83–S95. https://doi.org/10.1086/657241

Morera, Braulio. (2017). Planning new towns in the people's republic: The political dimensions of eco-city images in China. In Ayona Datta & Abdul Shaban (Eds.), *Mega-urbanization in the global south: Fast cities and new urban utopias of the postcolonial state* (pp. 188–204). Routledge.

Moser, Sarah. (2018). Forest city, Malaysia, and Chinese expansionism. *Urban Geography, 39*(6), 935–943. https://doi.org/10.1080/02723638.2017.1405691

Olds, Kris, & Yeung, Henry. (2004). Pathways to global city formation: A view from the developmental city of Singapore. *Review of International Political Economy, 11*(3), 489–521. https://doi.org/10.1080/0969229042000252873

Ong, Aihwa. (2011). Worlding cities, or the art of being global. In Ananya Roy & Aihwa Ong (Eds.), *Worlding cities: Asian experiments and the art of being global* (pp. 1–26). Blackwell.

Peck, Jamie. (2011). Geographies of policy: From transfer-diffusion to mobility-mutation. *Progress in Human Geography, 35*(6), 773–797. https://doi.org/10.1177/0309132510394010

Peck, Jamie, & Theodore, Nik. (2010). Mobilizing policy: Models, methods, and mutations. *Geoforum, 41*(2), 169–174. https://doi.org/10.1016/j.geoforum.2010.01.002

Pereira, Alex. (2002). The Suzhou industrial park project (1994–2001): The failure of a development strategy. *Asian Journal of Political Science, 10*(2), 122–142. https://doi.org/10.1080/02185370208434213

Perrons, Diane, & Posocco, Silvia. (2009). Globalising failures. *Geoforum, 40*(2), 131–135. https://doi.org/10.1016/j.geoforum.2008.12.001

Phelps, Nicholas. (2007). Gaining from globalization? State extraterritoriality and domestic economic impacts — The case of Singapore. *Economic Geography, 83*(4), 371–393. https://doi.org/10.1111/j.1944-8287.2007.tb00379.x

Pow, Choon-Piew. (2014). License to travel: Policy assemblage and the "Singapore model". *City, 18*(3), 287–306. https://doi.org/10.1080/13604813.2014.908515

Pow, Choon-Piew, & Neo, Harvey. (2015). Modelling green urbanism in China. *Area, 47*(2), 132–140. https://doi.org/10.1111/area.12128

Schäfer, Susann. (2017). The role of organizational culture in policy mobilities: The case of South Korean climate change adaptation policies. *Geografica Helvetica, 72*(3), 341–350. https://doi.org/10.5194/gh-72-341-2017

Shatkin, Gavin. (2014). Reinterpreting the meaning of the "Singapore Model": State capitalism and urban planning. *International Journal of Urban and Regional Research, 38*(1), 116–137. https://doi.org/10.1111/1468-2427.12095

Shatkin, Gavin. (2019). The planning of Asia's mega-conurbations: Contradiction and contestation in extended urbanization. *International Planning Studies, 24*(1), 68–80. https://doi.org/10.1080/13563475.2018.1524290

Sheppard, Eric. (2002). The spaces and times of globalization: Place, scale, networks, and positionality. *Economic Geography, 78*(3), 307–330. https://doi.org/10.2307/4140812

Singh, Bilveer (1988). Singapore: Foreign policy imperatives of a small state. Center for Advanced Studies, occasional paper. National University of Singapore.

Sioh, Maureen. (2010). The Hollow Within: Anxiety and performing postcolonial financial policies. *Third World Quarterly, 31*(4), 581–597. https://doi.org/10.1080/01436591003701109

Surbana. (2014). *Red-dotting the world.* Surbana International Consultants Pte Ltd.

SSTEC. (2019). *Who we are.* Retrieved from:https://www.mnd.gov.sg/tianjinecocity

Temenos, Cristina, & McCann, Eugene. (2013). Geographies of policy mobilities. *Geography Compass, 7*(5), 344–357. https://doi.org/10.1111/gec3.12063

The Economist (2012). *Africa's Singapore?.* The Economist Group. http://www.economist.com/node/21548263

The Straits Times. (1997). Jiang stresses close cooperation on Suzhou park project. Singapore Press Holdings.

Upadhya, Carol. (2020). Assembling Amaravati: Speculative accumulation in a new Indian city. *Economy and Society, 49*(1), 141–169. https://doi.org/10.1080/03085147.2019.1690257

Wells, Katie. (2014). Policyfailing: The case of public property disposal in Washington, D.C. *ACME: An International E-Journal for Critical Geographies, 13*(3), 473–494. https://acme-journal.org/index.php/acme/article/view/1023

Wilkinson, Iain. (2001). *Anxiety in a risk society.* Routledge.

Wong, Tai-Chee, & Goldblum, Charles. (2000). The China-Singapore Suzhou industrial park: A turnkey product of Singapore? *Geographical Review, 90*(1), 112–122. https://doi.org/10.2307/216177

Wu, Fulong. (2016). China's emergent city-region governance: A new form of state spatial selectivity through state-orchestrated rescaling. *International Journal of Urban and Regional Research, 40*(6), 1134–1151. https://doi.org/10.1111/1468-2427.12437

Yeo, George. (1997, August 4). A blunt talk with Singapore's Lee Kuan Yew: He's a model for China and the new Hong Kong. *Fortune.* https://money.cnn.com/magazines/fortune/fortune_archive/1997/08/04/229722/index.htm

Zhang, Jun. (2012). From Hong Kong's capitalist fundamentals to Singapore's authoritarian governance: The policy mobility of neo-liberalising Shenzhen, China. *Urban Studies, 49*(13), 2853–2871. https://doi.org/10.1177/0042098012452455

Conclusion

John Lauermann and Cristina Temenos

Failure matters. While stories of urban policy success are celebrated in the media and through professional networks, through an emphasis on 'learning from best practice' or 'knowledge translation', urban policy failures are just as likely, if not more likely, to occur. Indeed, policy failures are everywhere whether due to the realpolitik of internal debates, changing governance structures and ideologies, flaws in the conception or implementation, or because material realities have changed and existing policies are no longer fit for purpose. Scholarship within critical urban geography, and critical social sciences more broadly, often addresses failures in one form or another. However, these studies have not always sought to theorize failure directly, as the authors in this volume have done. The contributions to this book have highlighted the importance of paying heed to the processes and politics of policy failure. They demonstrate the value of thinking beyond "successism" (McCann and Ward 2015) in urban policy research. They provide a way into thinking through the material and political consequences of failure, and indeed the very nature of failure itself.

Yet as Mark Davidson argues in Chapter 1, this emerging research field must "face the question of what we want to know about policy failures." There are distinct epistemic communities researching the issue, each with their own priorities and foci. In an emerging literature, eclecticism should be embraced because each community contributes pieces that enrich the broader whole. Political science and public policy scholarship, for example, has focused on diagnosing why policies fail, often with detailed analysis of the design and management of the policies themselves (Dunlop 2017; Howlett, Ramesh, and Wu 2015; McConnell 2010). That research offers important insights into institutional process and frameworks for minimizing and ameliorating policy failures.

Geography and urban studies scholarship, in contrast, has been more concerned with failure as a political – and spatial – process. This research is attuned to a broader array of outcomes including, to borrow from Tom Baker and Eugene McCann in Chapter 4, "differentiation, mutation, fragility, unraveling, instability, emergence, detour, redirection, reaction, rejection, de-activation, and absence." The approach seeks to unravel the spatial dialectics of political action and policy failure, what Katie Wells describes in Chapter 2 as a messy combination of "practices, processes, and contexts through which a policy is made to fail," and through which failures are made political. For researchers like those represented in the book, failure is not only an unintended outcome. We are also interested in how policy failures are made to happen, for example through opposition, contestation, austerity, loss of support, changing priorities, sabotage, retribution, or neglect.

Throughout the book, the authors share a sense that something more substantial is at stake. Policy failure is not only about the policy, but also about the process and legacies of the failure. Failed policies do not just disappear. They linger past their expiration date, they fester in institutional memory, they leave scars on the landscape. Failed policy ideas get recycled into "new" policy agendas, sometimes demonstrating remarkable resilience over time and across changes in urban regimes. This volume highlights three specific reasons for this.

First, as demonstrated by Davidson in Chapter 1, Wells in Chapter 2, and Nciri and Levenda in Chapter 3, the definitions of "failure" and "success" are fluid within urban politics. As Nciri and Levenda argue, "politics of urban experiments are deeply intertwined with the social and power-laden construction of success and failure (and who defines them as such), and that actors can mobilize lessons from experimentation to slow down or block [policy goals]." There is a discursive as well as material function of failure. And in practice, policy is always portrayed as a success or failure by function of who is speaking and what their agenda is, belying the political nature of the terms themselves (Jessop 2011: Lorne 2021). What counts as failure and what counts as success is itself an outcome of politics (Holden, Scerri, and Esfahani 2015; Landau 2021). Those definitions are discursively and politically negotiated by competing stakeholders, by the changing of urban regimes, or even over the lifecycle of a policy project. Second, policy failure is not a static event that happens within a delimited instance. Things do not simply move on after a moment of failure. Rather, it is a relational and often ongoing process. It encompasses long social, political, and economic histories of particular places. It reflects how those places have developed in relation to other places, people, and events over time. This theme is taken up by Baker and McCann in Chapter 4, Moore-Cherry and Bonnin in Chapter 5, and Bok in Chapter 6. Moore-Cherry and Bonnin note in Chapter 5 that: "The city is an arena where different pasts, presents and futures come into friction with one another." Unpicking the ways in which these shifting temporalities converge in the city belies how failure is experienced differently by different stakeholders. Not all failure is equal, nor is it equally experienced across communities and places.

Third, as exemplified throughout all the chapters, policy failures often have generative effects. Positioning policy as somehow 'failed' opens political and material spaces to change, allowing new policies to replace the old, and the reworking of spaces affected by those policies for use in new, altered forms. Policy failure can also spur calls to action, being used in such a way as to motivate change, whether that is through new political actors coming into power, delegitimizing existing stakeholders, or bringing new ones into discussions on urban policy. Policy failure is then often about movement. Movement through spaces of urban governance, but also spaces of urban life, the city itself.

Landscapes of policy failure

All policy failures happen somewhere, of course. And it is no surprise that failures occur more frequently in some places than in others. The authors here have presented various approaches for conceptualizing the spatial dimensions of policy failure, though there is much work still to be done to analyze the reasons for geographic variation. Building on these individual cases, we might begin to develop a framework for incorporating failures within spatial theory.

That framework starts from the premise that policy failure has a spatially uneven geography. The unevenness likely reflects underlying geographic patterns and institutional arrangements – uneven levels of development, or spatial inequalities along socioeconomic lines, for instance. At a broad scale, there are differential rates of failure across cities, and also in the spaces of policy translation between cities. Policy mobilities research, for example, has demonstrated failed attempts at emulating policy models from elsewhere (Chang 2017; Müller 2015), failed attempts to export successful models to elsewhere (Stein et al. 2017; Malone 2018), partial failures (Wolfe 2016), and the troubling mobility of negative lessons and "worst practices" (Lovell 2019).

At a finer scale, policy failure is also unevenly distributed within metropolitan regions (Fricke 2020). Research on failing infrastructure in US cities, for example, has argued its geographic concentration within low income and minority neighborhoods is no accident. The uneven geographies of infrastructure failure instead reflect the spatial legacies of institutional and environmental racism, from redlining to NIMBYism (Cashin 2021; Kitzmiller and Drake Rodriguez 2021). One particularly jarring example is a 2019 sewer failure in a majority-Black neighborhood of New York City. Poorly designed and undermaintained infrastructure (much of it originally built under Robert Moses' deeply flawed highway projects) flooded 127 homes in Queens with raw sewage. Rather than remedying the failure, city officials blamed the victims for backing up sewers and used bureaucratic technicalities to reduce and delay payment on homeowners' claims – leaving many unable to repair their uninhabitable homes (Van Syckle 2021). It is almost unimaginable that similar failures would be allowed to occur in affluent neighborhoods of Manhattan, just a few miles away, even though the very same institutions manage the very same infrastructure network.

Empirically, the challenge is to analyze connections between general spatial inequalities and particular policy failures. This is not about assigning blame. Many policy failures have no single cause. They emerge instead from the cumulative effects of political conflicts or institutional missteps, and can sometimes be years or even decades in the making. Yet we do need to explain cause and effect if we are ever to correct failures of the past or prevent failures of the future. And it is important to understand how the decisions that lead to failures are made through political regimes, whether at municipal, regional, or national scales. Decision-making processes, from leadership down through everyday institutional practices, need to be analyzed in order to understand how the relationships between power, space, and community are differentially experienced and materialized. At minimum, critical research should seek to explain the unevenly distributed costs – and benefits – of policy failures.

Theoretically, the challenge is to interpret the city as a relational geography of policy outcomes, including policy successes, policy failures, and policy mediocrity too (Massey 2011). This means interpreting urban governance along a continuum, ranging from successful on one end to failed on the other, and acknowledging that most policies actually fall somewhere in between. For example, it is entirely possible for an urban policy regime to succeed in some categories (e.g. policing) while failing in others (e.g. schools), and even for those uneven successes and failures to work in tandem.

This also means policy geographies should be viewed as a patchwork, a mosaic of projects on the landscape, past and present, successful and failed. Patchwork geographies invoke their often ad hoc nature, or as Adey et al. (2021, 87) note, the work "that happens at multiple scales in ways that are often not joined up or cohesive within broader policy regimes." Therefore, the labor that goes into making policy and implementing it becomes an important point of analysis for thinking through policy success and failure. For example, an infrastructure network can reflect a multitude of policy decisions – ranging from wise to foolish – and a multitude of policy outcomes – ranging from wildly successful to abjectly failed. The decisions and outcomes have material consequences and are made up of the politics and processes that go into implementing policy. There are then a range of actors laboring to construct, advocate, oppose, and implement policy decisions that extend beyond politicians or technocrats that must be attended to in the analysis of policy failure (Baker, McCann, and Temenos 2020; De Coss-Corzo 2021). The geographic variations within the patchwork are, of course, correlated to uneven development patterns and socio-spatial inequalities. However, a patchwork perspective also attends to the gaps in the evolving nature of policy geographies, allowing space to think through transformational rather than iterative 'fixes' to policy failure.

From chronic condition to political action

Our introduction was originally written at the beginning of the COVID pandemic. We highlighted then how the pandemic exposed urban vulnerability, building on the cumulative effects of historical and ongoing policy failures. Several years on, policy failures remain salient as cities struggle to recover from the economic and social effects of the pandemic, which has lingered for longer than anyone had anticipated, and it is likely to continue on longer than anyone would hope. Like all crises, the pandemic brought to the fore underlying social truths. And just as with previous crises, the more consequential story looms in the background, for example the Great Recession revealing systemic flaws in the global financial system that disadvantages large swaths of the population.

The broader lesson of the pandemic – and as articulated by the growing urban policy failure literature – is that of chronic underperformance among urban institutions. Examples abound, ranging from dilapidated infrastructure (Millington and Scheba 2021) to the housing crisis (Fields and Hodkinson 2018; Wetzstein 2021). In each case, urban institutions persistently and continually fail to perform effectively. Some fail spectacularly, as seen in struggles to address the pandemic by woefully underfunded and underprepared public health departments. But more often, urban institutions fail gradually. They meander along, sufficiently managing but never actually fixing urban problems, such as the decades-long failure of school districts to provide quality education in low income, minority neighborhoods (Kitzmiller and Drake Rodriguez 2021). This kind of institutional mediocrity and low-grade political dysfunction is so common that it recedes from public view. Rather than generating a sense of political urgency, persistent and widespread failure overwhelms our political imagination, until we begrudgingly accept it as merely the backdrop of modern urban life.

The specific reasons for this state of affairs merit further research. This also begs the question addressed throughout this volume, 'failure for whom?' It is clear that the persistent yet incomplete nature of policy failures does not necessarily mean failure for everyone. There are large classes of people who benefit from certain kinds of policy failure. Surely half a century of "roll back neoliberalism" (Peck and Tickell 2002) has reduced institutional capacity through austerity. Extreme inequality (Dorling 2019), elite-led class war

(Harvey 2005), ideological denialism (Zizek 2011), and the willfully obtuse post-politics of technocratic leaders (Rancière 2006) have all eroded social solidarity to dangerously thin proportions.

And yet, through the pandemic as well as through earlier crises, new and different solidarities have been forged among and between communities (Karaliotas and Kapsali 2021; Nelson 2020). These new solidarities have sometimes come about in unexpected ways. While mutual aid cannot replace failures of the state, attention to more hopeful geographies can address the more specific causes and forms of policy failure more readily. In the process, these solidarities create space for new forms of planning, participation, or democracy (Elwood and Lawson 2020; Legacy 2021). Within the broad contours of capitalist crisis, there are often specific political, institutional, and even personal reasons why some policies fail while other succeed. These reasons are not necessarily easy to address, but they are at least possible to diagnose and contest. Persistent failure is neither inevitable nor desirable. More democratic, equitable, and just forms of urban governance are possible.

References

Adey, Peter, Tim Cresswell, Jane Yeonjae Lee, Anna Nikolaeva, André Nóvoa, and Cristina Temenos. 2021. *Moving Towards Transition: Commoning Mobility for a Low-Carbon Future*. Bloomsbury Publishing.

Baker, Tom, Eugene McCann, and Cristina Temenos. 2020. "Into the ordinary: Non-elite actors and the mobility of harm reduction policies." *Policy and Society* 39 (1):129–145. doi: 10.1080/14494035.2019.1626079.

Cashin, Sheryll. 2021. *White Space, Black Hood: Opportunity Hoarding and Segregation in the Age of Inequality*. Beacon Press.

Chang, I-Chun Catherine. 2017. "Failure matters: Reassembling eco-urbanism in a globalizing China." *Environment and Planning A: Economy and Space* 49 (8):1719–1742. doi: 10.1177/0308518x16685092.

De Coss-Corzo, Alejandro. 2021. "Patchwork: Repair labor and the logic of infrastructure adaptation in Mexico City." *Environment and Planning D: Society and Space* 39 (2):237–253. doi: 10.1177/0263775820938057.

Dorling, Danny. 2019. *Inequality and the 1%*. 3rd ed. London: Verso.

Dunlop, Claire A. 2017. "Policy learning and policy failure: Definitions, dimensions and intersections." *Policy & Politics* 45 (1):3–18. doi: 10.1332/030557316X14824871742750.

Elwood, Sarah, and Victoria Lawson. 2020. "The arts of poverty politics: Real change." *Social & Cultural Geography* 21 (5):579–601. doi: 10.1080/14649365.2018.1509111.

Fields, Desiree J., and Stuart N. Hodkinson. 2018. "Housing policy in crisis: An international perspective." *Housing Policy Debate* 28 (1):1–5. doi: 10.1080/10511482.2018.1395988.

Fricke, Carola. 2020. "Implications of metropolitan policy mobility: Tracing the relevance of travelling ideas for metropolitan regions." In *Metropolitan Regions, Planning and Governance*, edited by Karsten Zimmermann, Daniel Galland and John Harrison, 117–132. Springer.

Harvey, David. 2005. *A Brief History of Neoliberalism*. Oxford University Press.

Holden, Meg, Andy Scerri, and Azadeh Hadizadeh Esfahani. 2015. "Justifying redevelopment 'failures' within urban success stories': Dispute, compromise, and a new test of urbanity." *International Journal of Urban and Regional Research* 39 (3):451–470. doi: 10.1111/1468-2427.12182.

Howlett, Michael, M Ramesh, and Xun Wu. 2015. "Understanding the persistence of policy failures: The role of politics, governance and uncertainty." *Public Policy and Administration* 30 (3–4):209–220. doi: 10.1177/0952076715593139.

Jessop, Bob. 2011. "Metagovernance." In *The SAGE Handbook of Governance*, 106–123. SAGE.

Karaliotas, Lazaros, and Matina Kapsali. 2021. "Equals in solidarity: Orfanotrofio's housing squat as a site for political subjectification across differences amid the 'Greek crisis.'" *Antipode* 53 (2):399–421. doi: https://doi.org/10.1111/anti.12653.

Kitzmiller, Erika M., and Akira Drake Rodriguez. 2021. "Addressing our nation's toxic school infrastructure in the wake of COVID-19." *Educational Researcher*. doi: 10.3102/0013189x211062846.

Landau, Friederike. 2021. "Agonistic failures: Following policy conflicts in Berlin's urban cultural politics." *Urban Studies* 58 (12):2531–2548. doi: 10.1177/0042098020949080.

Legacy, Crystal. 2021. "The point is still to change it." *Planning Theory & Practice* 22 (4):511–515. doi: 10.1080/14649357.2021.1962054.

Lorne, Colin. 2021. "Struggling with the state I am in: Researching policy failures and the English National Health Service." *Emotion, Space and Society* 38:100746. doi: https://doi.org/10.1016/j.emospa.2020.100746.

Lovell, Heather. 2019. "Policy failure mobilities." *Progress in Human Geography* 43 (1):46–63. doi: 10.1177/0309132517734074.

Malone, Aaron. 2018. "(Im)mobile and (Un)successful? A policy mobilities approach to New Orleans's residential security taxing districts." *Environment and Planning C: Politics and Space*. doi: 10.1177/2399654418779822.

Massey, Doreen. 2011. "A counterhegemonic relationality of place." In *Mobile Urbanism: Cities and Policymaking in the Global Age*, edited by Eugene McCann and Kevin Ward, 1–14. University of Minnesota Press.

McCann, Eugene, and Kevin Ward. 2015. "Thinking through dualisms in urban policy mobilities." *International Journal of Urban and Regional Research* 39 (4):828–830. doi: 10.1111/1468-2427.12254.

McConnell, Allan. 2010. "Policy success, policy failure and grey areas in-between." *Journal of Public Policy* 30 (3):345–362. doi: 10.1017/S0143814X10000152.

Millington, Nate, and Suraya Scheba. 2021. "Day zero and the infrastructures of climate change: Water governance, inequality, and infrastructural politics in Cape Town's water crisis." *International Journal of Urban and Regional Research* 45 (1):116–132. doi: https://doi.org/10.1111/1468-2427.12899.

Müller, Martin. 2015. "(Im-)Mobile policies: Why sustainability went wrong in the 2014 Olympics in Sochi." *European Urban and Regional Studies* 22 (2):191–209. doi: 10.1177/0969776414523801.

Nelson, Anitra. 2020. "COVID-19: Capitalist and postcapitalist perspectives." *Human Geography* 13 (3):305–309. doi: 10.1177/1942778620937122.

Peck, Jamie, and Adam Tickell. 2002. "Neoliberalizing space." *Antipode* 34 (3):380–404.

Rancière, Jacques. 2006. *Hatred of Democracy*. Translated by Steve Corcoran. Verso.

Stein, Christian, Boris Michel, Georg Glasze, and Robert Pütz. 2017. "Learning from failed policy mobilities: Contradictions, resistances and unintended outcomes in the transfer of "Business Improvement Districts" to Germany." *European Urban and Regional Studies* 24 (1):35–49. doi: 10.1177/0969776415596797.

Van Syckle, Katie. 2021. "Flooded with sewage, and still no help." *The New York Times*, April 16, A16.

Wetzstein, Steffen. 2021. "Toward affordable cities? Critically exploring the market-based housing supply policy proposition." *Housing Policy Debate*, 1–27. doi: 10.1080/10511482.2021.1871932.

Wolfe, Sven Daniel. 2016. "A silver medal project: The partial success of Russia's soft power in Sochi 2014." *Annals of Leisure Research* 19 (4):481–496. doi: 10.1080/11745398.2015.1122534.

Zizek, Slavoj. 2011. *Living in the End Times*: Verso Books.

Index

Note: Numbers with 'n' indicate notes in the text.